Water Utility Capital Financing

AWWA MANUAL M29

Second Edition

FOUNDED
1881

American V

MANUAL OF WATER SUPPLY PRACTICES—M29, Second Edition

Water Utility Capital Financing

Project Manager and Technical Editor: Kathleen Faller
Production Editor: Alan Livingston

Library of Congress Cataloging-in-Publication Data

Water utility capital financing -- 2nd ed.
xiii, 87 p. 17×25 cm. -- (AWWA manual ; M29)
Includes bibliographical references and index.
ISBN 0-89867-957-5
1. Municipal water supply--United States--Finance--Handbooks, manuals, etc. 2. Waterworks--United States--Finance--Handbooks, manuals, etc. I. American Water Works Association. II. Series.
TD491.A49 no. M29 1998
628.1 s--dc21
[363.6'1'0681] 98-16938
CIP

Printed in the United States of America
American Water Works Association
6666 West Quincy Avenue
Denver, CO 80235

ISBN 0-89867-957-5

Printed on recycled paper

Contents

Tables

Figures

Acknowledgments

The AWWA Management Division Board of Trustees gratefully acknowledges the contributions made by members of the Finance, Accounting, and Management Controls Subcommittee—and in particular the M29 Working Group—in preparing this manual. The subcommittee's efforts aid in meeting the general need for continued improvement in financial planning by water utilities for their ongoing capital requirements.

The board also expresses appreciation to those past and present members of the committee who submitted information, drafted text, and submitted editorial comments for preparing the manuscript. The Management Division thanks the entire subcommittee for its time, expertise, and exchange of views, which have resulted in this comprehensive revision of the financial management manual.

Finance, Accounting, and Management Controls Subcommittee

R. Miller, Chair, Louisville Water Company, Louisville, Ky.
T.R. Barbee, Co-chair, Arlington Water Department, Arlington, Texas*
K.M. Corry, Co-chair M29 Working Group, Denver Water, Denver, Colo.*
R.D. Giardina, Co-chair M29 Working Group, Rick Giardina & Associates, Inc., Aurora, Colo.*
B. Brunwasser, Philadelphia Water Department, Philadelphia, Pa.
J.W. Fagan , Northbridge Environmental Consultants, Vienna, Va.*
J.S. Feld, New York City Municipal Water Finance Authority, New York, N.Y.
K.A. Gibson, Contra Costa Water District, Concord, Calif.
J.J. Hansen , Bridgeport Hydraulic Company, Bridgeport, Conn.*
J.S. Jerrell, South Central Connecticut Regional Water Authority, New Haven, Conn.*
D.A. Kaplan, City of Philadelphia Water Department, Philadelphia, Pa.*
J.S. Kelly, City of Madison, Madison, Wis.
D.D. Lanning, Camp Dresser & McKee, Cambridge, Mass.*
S. Lehman, Standard & Poor's Rating Group, New York, N.Y.*
K.P. Mulshine, Prager, McCarthy & Sealy, Orlando, Fla.*
J.M. Mussman, Black & Veatch, Gaithersburg, Md.*
G.H. Nickerson Jr., Nabors, Giblin and Nickerson LLC, Tallahassee, Fla.
T.C. Schlette, Malcolm Pirnie, Inc., White Plains, N.Y.*
L.G. Shaw, Dallas, Texas
Z.L. Snyder, Black & Veatch, Gaithersburg, Md.*
R. Thomas, Wyoming Michigan Water District, Wyoming, Mich.

Other Participants

D. Segal, Nations Bank, Charlotte, N.C.*

*Denotes M29 Working Group member

Introduction

The financial condition of many water utilities has been continually challenged because of a variety of factors, such as inflation, growth-related demands, and public pressure to keep water rates low. This trend, which began in the 1970s, is likely to continue through the close of this century and into the next. Despite growing maintenance needs, water utilities tend to postpone repairs, replacements, and minor improvements in order to keep operating costs and corresponding service charges as low as possible in the short term.

In recent years, the resulting infrastructure crisis in the United States has risen to the forefront of the nation's conscience. Problems with the nation's infrastructure stem from neglect and underinvestment in the repair and replacement of public capital facilities. Deferred maintenance expenditures result in higher rehabilitation costs. Even investor-owned water utilities suffer from regulatory changes and denial of some critical rate increases. Surveys indicate that investor-owned utilities' infrastructure problems are similar to those experienced by publicly owned utilities.

Along with electric, natural gas, stormwater, and wastewater utilities, water utilities remain one of the most capital-intensive business operations. As demands for new service collide with continuing stringent capital-related regulatory requirements, the need for well-conceived, rational capital funding plans cannot be overemphasized.

The American Water Works Association (AWWA) strongly advocates that utility management, governing bodies, and regulatory agencies address the infrastructure crisis by realizing that customers' interests are best served through well-maintained and well-operated water systems that are supported by funding sufficient to meet utility needs.

Adequate financing for infrastructure no longer means looking primarily to federal or state grant funding sources. An effective financial program may contain a variety of approaches to financing infrastructure. Government assistance, which for many utilities dominated capital funding plans of the 1970s and 1980s, currently exists only for select projects and on a much smaller scale than previously. Financial solutions for the remainder of the 1990s and into the next century may include well-structured public and private partnerships to meet infrastructure needs.

Thus, the subject of contemporary financing concepts is particularly relevant to the water utility industry. Many water utilities are faced periodically with increased requirements for capital funds in order to meet demands placed on their systems by rapid growth. Other utilities face rehabilitation and replacement needs. Whatever a particular water utility's capital requirements may be, knowledge of available alternatives for obtaining capital funds is a key element in developing successful financing plans.

PURPOSE

This manual provides a comprehensive overview of traditional and innovative financing sources that water utility management should evaluate when determining long- and short-term strategies for meeting capital requirements. Only by thorough analysis of a particular water utility's long-term capital requirements and financial position will the feasibility of one or more specific financing concepts be determined.

This manual is not intended to be, nor should it be considered, a complete text on water utility capital-financing concepts; a variety of other sources are available through various industry and professional organizations. Instead, the text should primarily be considered a guide for determining capital requirements, strategic planning, and financing alternatives. It also provides general information for those utilities considering the issuance of long-term debt. Each of these areas is critical to the capital-planning-and-financing process, but must be tailored to and reflect factors applicable to a local situation. No recommendations or opinions about the relevance of alternative forms of available financing have been provided.

SCOPE

The material presented in this manual is intended for use by both publicly owned and investor-owned water utilities. Because financing alternatives and markets may differ for these two primary types of utility ownership, financing alternatives for each are discussed in separate places in this manual.

Information presented in this manual has certain limitations. Water utility capital requirements continually change, as do financial sources available to the utility. At any given time, financing alternatives available to a utility depend on

1. Tax laws and Internal Revenue Service (IRS) rules and regulations.
2. Interest rates charged for various financing alternatives.
3. A utility's credit rating.
4. Decisions by governing bodies, by regulatory agencies, and, in some instances, by voters.

New financing alternatives are created and old ones abandoned as market conditions change. Some financing options may become obsolete until future market changes or legislative actions make them feasible again.

This manual covers six general areas: (1) strategic capital and financial requirements planning, (2) identifying financing alternatives, (3) evaluating financing alternatives, (4) the process of taking a debt issue to the market, (5) participants' roles in the debt issuance process, and (6) special considerations for investor-owned utilities. Additional information regarding the requirements of different rating agencies is contained in appendix A.

Proper planning of capital requirements begins with the preparation of a strategic plan that communicates utility management goals to personnel, customers, investors, and the community as a whole. The utility's annual capital budget and capital improvement program (CIP) reflect established goals and serve as short- and long-range planning tools. The strategic plan and resulting CIP are dynamic tools requiring periodic updating or revalidation.

A critical element in the strategic planning process is the development of a financial plan that integrates the CIP. Financial planning involves forecasting customer demand, water sales, revenues, and operating expenditures in order to develop an initial cash flow analysis that reflects current and prospective utility practices.

Alternative financing methods should be analyzed and compared to the initial plan to determine any cost savings and shared benefits between the ratepayers and the utility. Alternatives should be further evaluated for their ability to meet utility objectives and rating agency requirements.

Investor-owned systems have special considerations for meeting financing requirements. In addition to the normal business risks of attracting capital in a manner that maintains the utility's capital structure, investor-owned utilities must provide an adequate return on stockholders' investments and balance the needs for renewals and replacements through depreciation and equity funding. Regulatory agency decisions play an important part in the investor-owned utility's ability to meet its capital requirements. The potential effects on the utility and ratepayers of regulatory decisions regarding financing alternatives should be evaluated.

Another planning consideration consists of convincing local government policymakers, regulatory agencies, and rating agencies of the benefits of proposed capital plans. Well-documented strategic plans will assist in this activity. Also, because change is inevitable, sensitivity analyses of potential contingency situations should be evaluated for their impacts on utility planning. Such analyses will lead to better management decisions and avoid problems when real contingencies occur.

Appendix A provides information regarding the requirements of bond-rating agencies, which is particularly useful as a utility considers the use of long-term debt and begins the process of going to the financial market.

The glossary at the end of this manual, which provides terms commonly used by the financial community, is a continuing step toward establishing uniformity of definitions. It will be revised and expanded as additional AWWA financial and accounting manuals are developed.

AWWA advocates that specific financing requirements for any particular water utility be based on sound economic, financial, accounting, and engineering principles. Often, the services of consultants and counselors experienced in such matters are required. When such advisors are employed, this document should serve as a resource that the policymakers and utility managers may draw on to guide their evaluations of the validity of any proposed financing alternative.

This manual is one of several AWWA manuals and books that address financial management topics. Others that may interest the reader include: *Water Utility Accounting*, 3rd ed.; AWWA Manual M1, *Water Rates*; AWWA Manual M26, *Water Rates and Related Charges*; AWWA Manual M34, *Alternative Rates; Financial Planning Model for Water Utilities*, software; and AWWA Research Foundation's, *Meeting Future Financing Needs of Water Utilities*.

Chapter 1

The Capital Finance Planning Environment

The demands and expectations of water utility stakeholders, i.e., customers, regulators, and policymakers, create a challenging capital-financing planning environment for water utility management. In addition to demanding consistently safe and reliable drinking water, stakeholders expect water service to be provided efficiently, at minimum annual cost, and with minimal impact on ratepayers.

This chapter discusses the planning environment in which capital-financing decisions are made, as illustrated in Figure 1-1. The discussion emphasizes the importance of management comprehensively evaluating all assumptions used in capital program development. It is essential that management perform this evaluation as it moves from strategic planning, reviewing historical capital programs, and assessing financial performance to developing specific action plans in which projects are prioritized. Comprehensive capital program planning requires more than life-cycle costing. The financial impacts on utility operations and customers' bills must also be considered. Accordingly, utility management prudently integrates overall financial planning, revenue requirements planning, and rate-setting analyses into an iterative capital-financing planning process.

A fully integrated and cost-effective capital program and financial plan are rarely developed in a single attempt. Most utility managers prepare multiple planning scenarios before selecting a course of action to effectively address the outcomes expected by the utility's stakeholders. By their very nature, all plans must be considered living documents that are to be reviewed and updated regularly. The capital-financing planning environment water utilities face is one of challenge and continual change, requiring an integrated iterative planning process.

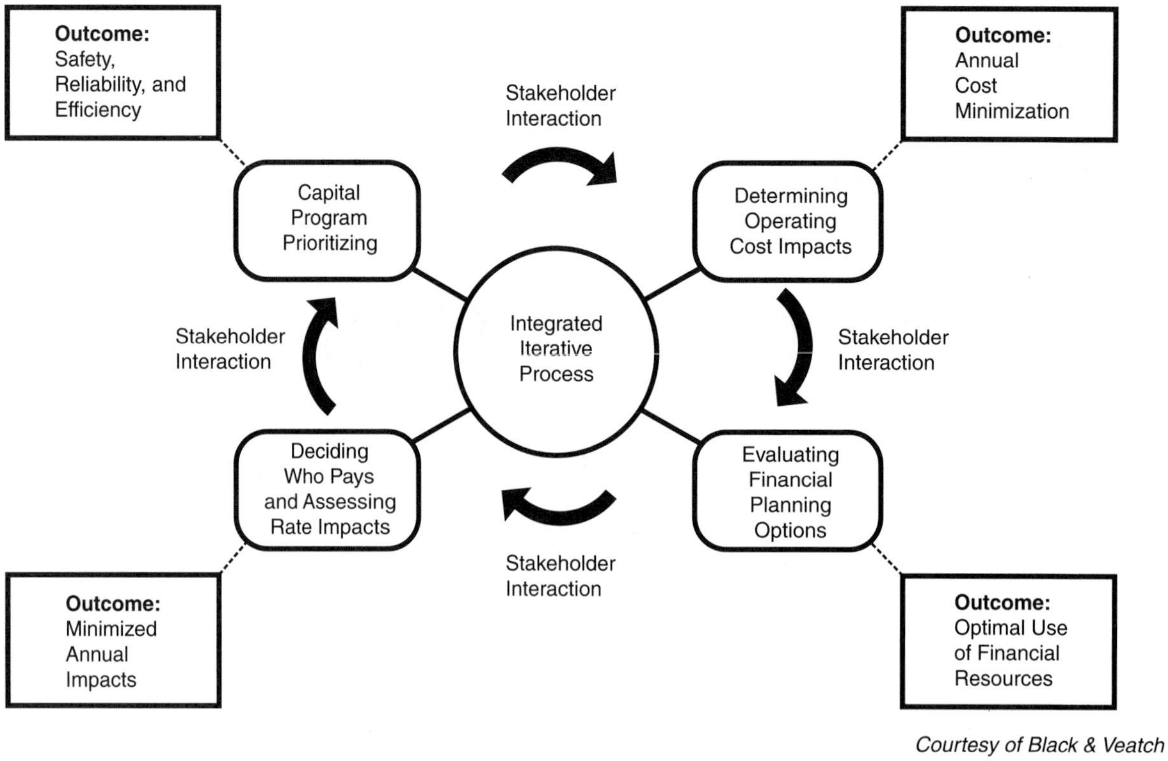

Figure 1-1 Financial planning flowchart

CAPITAL REQUIREMENTS STRATEGIC PLANNING

A strategic plan is a formal document that communicates management's goals for the future to personnel, customers, investors, and the community as a whole. The plan provides a basic approach and structure through which a utility develops (among other activities) appropriation requests that reflect decisions about how resources will be obtained and used.

Strategic planning represents a long-term look into the future operations of a utility. As part of this process, management must define the goals and objectives that map a path between a utility's present position and its vision of the future. This definition includes a description of the utility's strategies and tactics for accomplishing those goals and objectives. Strategic planning also explicitly defines outcomes and outputs expected by management, providing a gauge that permits performance monitoring. It serves as a guide for management, influencing future resource allocation and operational decisions. Strategic planning is a dynamic and continual process providing management with direction for all operations, including capital additions and replacements.

A successful strategic planning process provides many benefits to a utility and to those affected by its operations. Strategic planning improves a utility's ability to anticipate and accommodate future conditions by identifying issues, opportunities, and problems. Management uses its strategic plan to develop a comprehensive master plan. The master plan addresses the entity's future capital requirements stemming from growth in service demands, maintenance needs, and required replacements and improvements to the existing utility plant.

Planning Components

In preparing a strategic plan, management must focus on the utility's mission and core values, as well as the operating philosophy to be used to meet future goals. Thus, the components of strategic planning include

- a mission statement
- a statement of core values
- a description of the operating philosophy (goals, priorities, action plans, and progress indicators)

The mission statement describes a utility's basic reason for existence. This may be clearly stated by legislation, corporate certification, or local ordinance. Core values for a utility reflect the principles of conduct for carrying out the utility's mission. Finally, the operating philosophy defines the utility's goals, outlines key factors for evaluating success in achieving the mission and goals, and indicates the methods that will be used to implement strategies.

The strategic plan presents each of these components in comprehensive yet simply phrased statements. Starting with the mission statement, management narrows the focus to goals and then to detailed action plans. Each component of the strategic planning template shown in Figure 1-2 is guided by the previous one. At the fourth level, management defines its goals. Directed by the utility's mission statement and core values, management prepares stated goals for the utility. To reach these goals management then prepares detailed action plans for the future, generally ranging from 5 to 10 years.

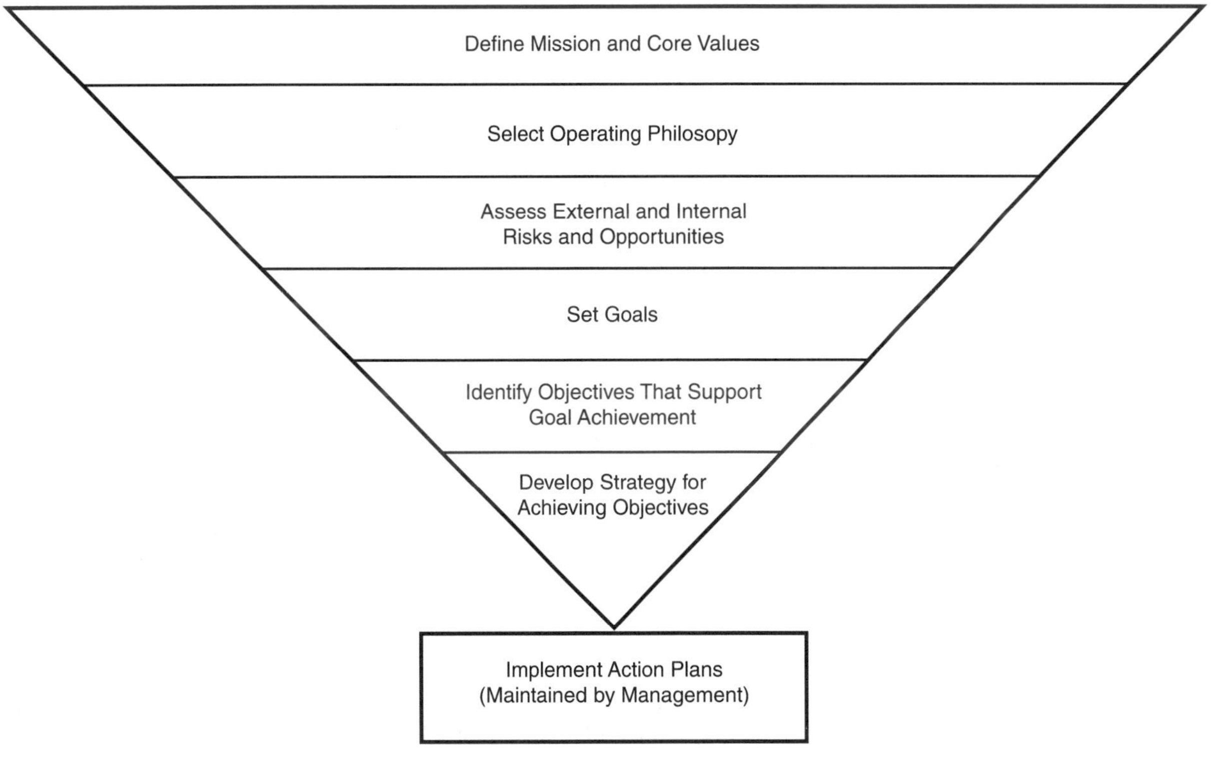

Source: Governor's Office of Budget and Planning, Texas

Figure 1-2 Strategic planning template

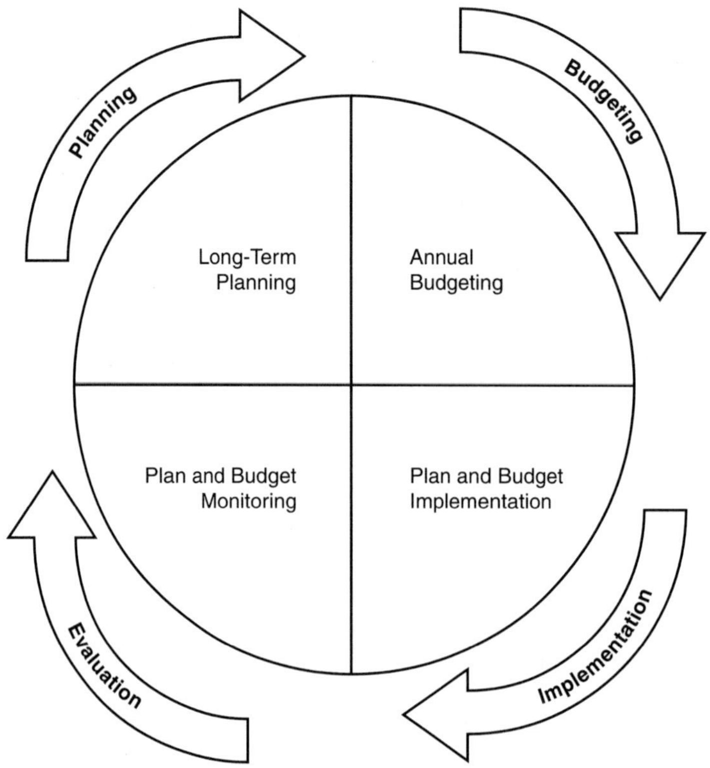

Source: Governor's Office of Budget and Planning, Texas

Figure 1-3 Strategic planning and budgeting process

Periodic assessments should be undertaken by management to measure both efficiency and effectiveness in reaching stated goals. Targets for specific actions must be determined during the planning phase, and the impacts of those actions must be measured during the implementation phase. In other words, the following questions should be addressed:

1. Are we doing what we planned to do?
2. Are we getting the results we intended?

A good assessment process should answer both of the preceding questions. Strategic plans must be fluid and should be updated as changes occur or when the action plans do not yield the desired results. Figure 1-3 demonstrates this iterative process. The Governmental Accounting Standards Board (GASB) encourages local and state governmental entities to experiment with performance measures and standards reporting to integrate performance measures into a utility's capital program. GASB is expected to require some level of performance measures and standards reporting around the turn of the century.

In the current environment, strategic planning is affected by numerous elements and assumptions. Some of these can be influenced by a utility's actions; others are beyond the utility's control. These elements include

- availability of source-of-supply waters
- federal and state regulatory mandates

- water system infrastructure conditions
- a utility's existing financial condition
- demographics and increases or decreases in customer demand
- economic conditions within the utility's service area

Utility resources are allocated to address these and other factors based on the utility's mission statement and on the goals designed to enhance the utility's ability to achieve its mission. Goals addressing each of the preceding considerations should be prepared. However, it is likely that some of these goals will conflict with others. Utility management must weigh the demands from each element and prioritize its goals. Lower-priority goals that conflict with other goals may receive little or none of a utility's resources.

The strategic plan provides direction for developing the utility's master plan and CIP. The CIP provides the forecast of evaluations, studies, facilities plans, and tangible capital costs for a multiyear period, generally 5 to 10 years or some other utility-designated planning period.

Historical Performance of CIP Implementation

One of the most common characteristics used to measure strategic planning achievements is a utility's ability to implement its adopted CIP. A utility develops a schedule for completion of CIP projects based on its goals, and the schedule becomes a fundamental part of the utility's action plan. Figure 1-4 shows a sample CIP schedule of project priorities resulting from a utility's strategic planning process. An adopted CIP has a schedule and budgeted cost based on management estimates, as detailed in later sections of this chapter. The strategic plan provides management with basic guidance for determining when each CIP project will be implemented and how much of the utility's resources can be allocated toward completion of projects at any given time. Success in meeting strategic planning goals can be measured by comparing the budgeted CIP to actual work conducted during a period of time.

Ideally, the CIP will be completed on schedule, demonstrating management's ability to achieve its stated goals. However, there are numerous reasons a CIP schedule might not be met; these causes should be identified and evaluated. If schedule slippage is within management's control, management can take action to correct this problem in the future during strategic planning efforts. If the schedule has slipped because assumptions used in developing the strategic plan have changed, the action plans (usually the utility master plan) may need reevaluation and change. Regardless of the reason for falling behind the CIP schedule, the utility needs to adjust its CIP to reflect its implementation capabilities and changes in its allocation of resources.

System Growth Rates

A fundamental consideration in determining a utility's future capital requirement (CIP) is the population growth of the area the utility serves. This information is essential to prepare demand estimates for a master plan and a CIP. Because obtaining creditable population forecast numbers can be difficult, management generally relies on a review of historical service area growth patterns as a guide for planning future capital requirements. Historical billing information for a 5- to 10-year period may be readily accessible and may provide a reasonable basis of growth trends from which to project the number of customers in each class of service. A comparison of census population figures with customer accounts may provide a reasonable correlation from which to equate population with future numbers of customers.

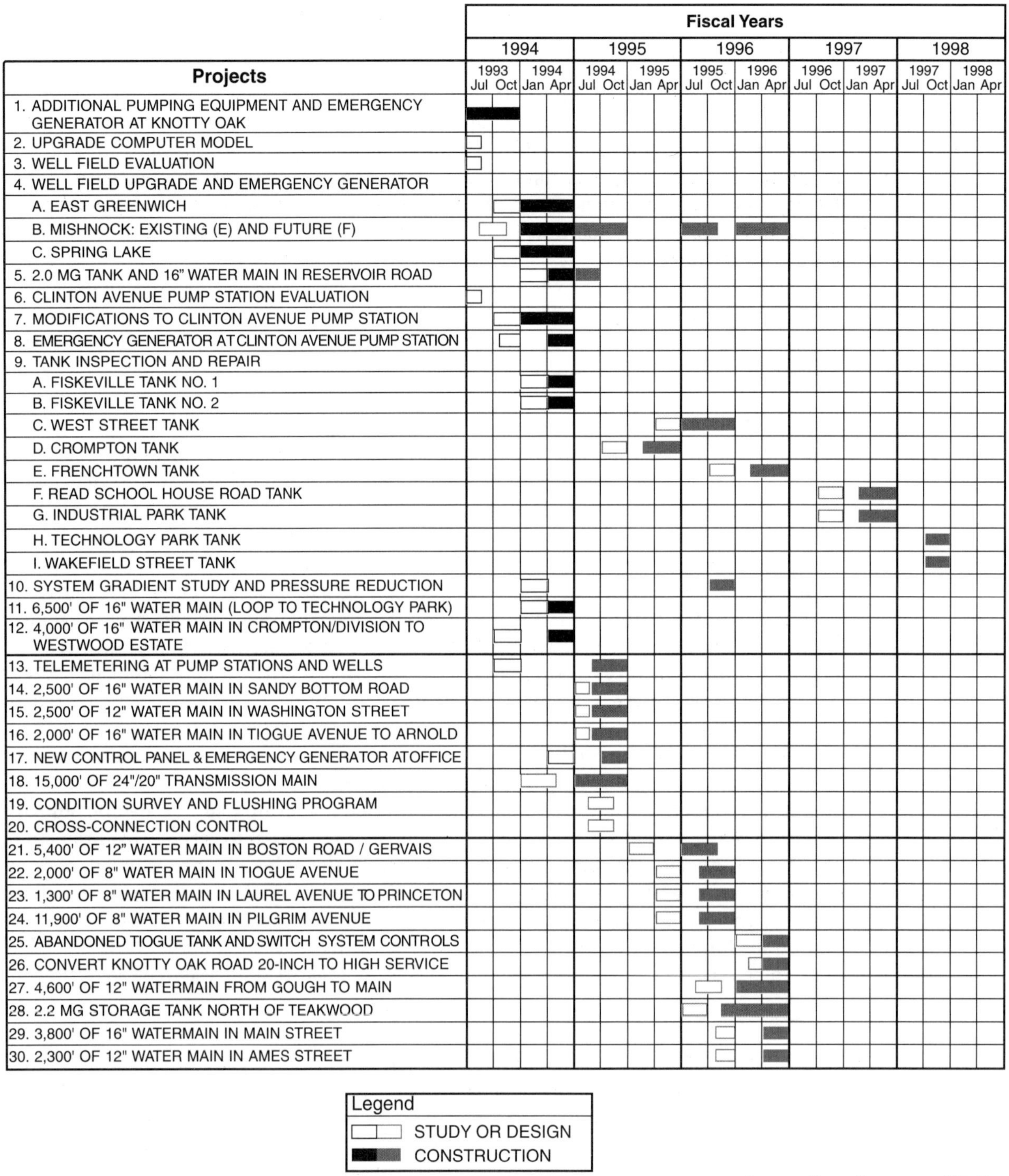

Source: Camp Dresser and McKee

Figure 1-4 Sample CIP schedule

Once the historical population growth trends are developed, projections can be made for each class of customer based on historical trends, known changes to the demographics specific to each class, and economic trends such as industrial and commercial expansions or shutdowns. This process includes an evaluation of local, county, and state projections for construction starts and for estimates of construction activity until build-out of the service area is achieved.

Conservatism Versus Optimism in Projections

It is typical for planning purposes to develop a range of estimates for management to consider. This range starts with a conservative plan assuming the worst-case population growth scenario and ends with a best-case growth scenario. Management must then select a baseline scenario for conducting the remaining portion of the planning process.

Worst and best cases can be interchangeable according to the priority of the considerations and assumptions used to develop a utility's strategic plan. For example, if new supply is a high priority because the utility does not have the capacity to serve additional demand, the worst-case scenario could be increased population growth and service demand. This scenario would require costly capital additions for supply, treatment, transmission, and distribution. Here, the best-case scenario could be slow population growth. On the other hand, if supply is not a high priority and the utility has excess capacity available, the worst case could be no population growth and the best case could be increased population or new wholesale customers outside the service area that will enable the utility to take advantage of economies of scale.

The acid test for any scenario is management's ability to demonstrate the probability that the chosen scenario will occur. In making the final presentation for budget approval, management must perform a "reality check" and choose one scenario for planning purposes. The chosen scenario should produce results representing the least degree of financial risk to the utility while ensuring that plant additions will be placed in service before capacity is required to meet increased demand.

Alternative scenarios may also be used in some situations, depending on the intended use. For example, growth and population projections suitable for facility-planning purposes may result in enhanced levels of risk for financial planning purposes. The goal of facility planning is to ensure that service demands can be met. Hence, increased population growth (worst-case) projections might be appropriate for facility-planning purposes. However, basing rates and revenue projections on the same population growth estimate may understate the actual required future rate levels and increase financial risk for the utility. Frequently, the use of two scenarios—one for facility-planning purposes and a second for financial planning—may be appropriate.

Customers' Service Needs

A utility generally has an obligation to serve all customers, both new and existing. The capital cost of providing service to new customers can be significantly higher than the embedded cost of serving existing customers. The strategic planning process should include a policy decision by management about cost recovery for growth-related additions to capacity, i.e., a determination of who pays for the cost of growth.

The answer to the question of who pays is not as clear as it would first appear. Not all additions to capacity benefit only new customers. Some plant additions may also improve the quality of service to existing customers. Service objectives for existing and new customers are in many cases the same: water quality and safety, as well as service reliability. However, current customers seek to minimize the impact of growth on general service charges, while new customers would like to minimize front-end costs to connect to the water system.

To the extent that capital additions satisfy similar objectives for both new and existing customers, the question of who pays becomes cloudy. However, capital costs that clearly benefit only new customers can be funded by those new customers directly through various means, including system development charges, as discussed in detail in AWWA Manual M26, *Water Rates and Related Charges*.

Integrated Resource Planning

A water utility must look at all available methods of optimizing its operational and economic resources. Integrated resource planning (IRP) is a process to maximize available resources by considering a wide range of supply-side and demand-side resources. The process provides information on possible consequences and helps management judge the value of trade-offs among resource alternatives.

IRP for water utilities has its roots in a method of planning used by electric and gas utilities in the 1980s when supply was short and the cost of additional supply became a significant factor. Basically, strategic planning using IRP looks not only at cost and availability of supply but also at a utility's ability to manage demand. When properly applied, the IRP process leads to sound long-term decisions and lower overall costs.

IRP brings a relatively new dimension into the strategic planning process for water utilities and opens the door to explicit evaluation of options that were not previously considered, such as wastewater reuse, demand management, and planned shortages. Figures 1-5 and 1-6 illustrate the trade-offs IRP presents that water utilities have not generally analyzed in the past. An IRP process should result in a common perception by both the utility and the public as to what is acceptable for cost and levels of reliability. Figure 1-6 illustrates two possible different cases for the same level of reliability, both of which are related to the investment policy of the utility.

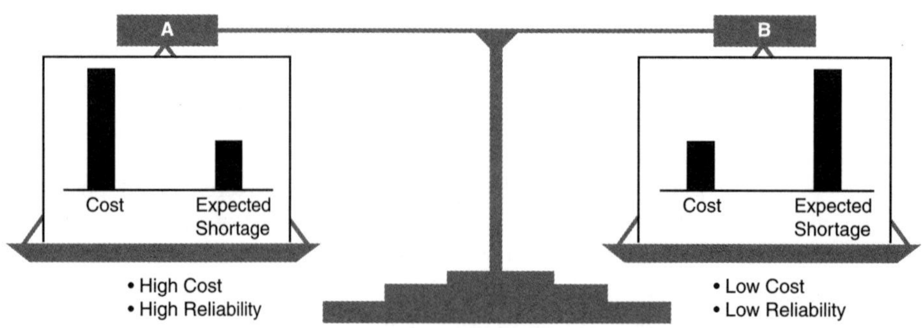

Source: E.F. Brigham

Figure 1-5 Trade-offs between reliability and cost

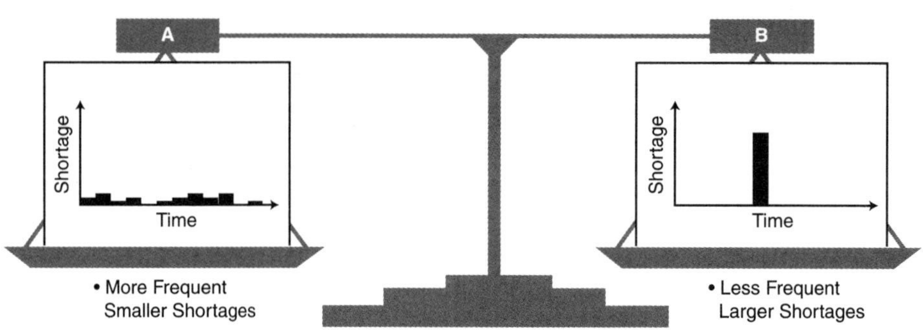

Source: E.F. Brigham

Figure 1-6 Different aspects of reliability

System Master Plan and Strategic Capital Requirements

A completed strategic plan enables a utility to develop a master plan of capital requirements and to determine funding sources needed to complete those capital projects. The master plan provides the following benefits:

- It summarizes major goals and objectives.
- It characterizes the existing system.
- It defines utility needs.
- It presents capital requirements for fulfilling the needs assessment.

From this master plan, the utility will develop a specific CIP. Both capital planning and the CIP are addressed in the next section.

CAPITAL BUDGETING

Within the framework of major objectives and policies established for a water utility, management must plan and control capital expenditures. Planning encompasses long- and short-term anticipation of customer needs for water service, selection and design of proper plant facilities to meet customer needs, timely purchase or construction of required plant additions, and timely replacement and retirement of plant assets. The capital plan should be prepared for a period of up to 10 years. In some cases, the capital plan may tie into a master plan of 20 years or more duration. The resulting multiyear capital expenditures plan should reflect management's understanding of strategic objectives.

A capital budget is a 1-year plan that covers additions, improvements, and replacements to an existing utility system. This annual budget could be separated into major and routine capital projects. Major capital projects include proposed additions and improvements that involve relatively large expenditures and are of a nonrecurring nature. Routine capital projects include proposed additions, replacements, and improvements involving relatively small expenditures that generally recur on an annual basis. The annual capital budget can be used to systematically implement the long-range plan.

Utility management should use its strategic plan to guide annual budgeting and capital expenditure requirements, periodically reviewing such factors as financial, water resources, system design, growth restrictions, infrastructure rehabilitation, and changing customer service requirements.

Regulations

When planning capital requirements, utility management should be aware of the requirements and dictates of all regulatory agencies at federal, state, and local levels. For investor-owned water utilities, regulatory control is usually exercised by a state regulatory agency. The authority given to these agencies is generally comprehensive. Publicly owned utilities are generally regulated by local governments. In some states, publicly owned utilities or those serving across political boundaries are regulated by state agencies. A utility should take positive steps to determine whether it is in compliance with these regulatory agencies or to verify that allowances in capital and operating budgets have been made to address deficiencies that become evident as a result of specific regulations. For example, legislation such as the Safe Drinking Water Act has caused many utilities to make major modifications to capital plans. In general, the long-term CIP needs to recognize the effects of current and proposed legislation.

Developing the Capital Plan

The process of developing a multiyear capital plan should integrate all aspects of planning, so that not only are the priorities and timing of new facilities determined, but the impacts on operating costs are also estimated. The effect of the capital plan on human resources and material requirements serves as the basis for planning the operation and construction activities the utility will perform. Through this comprehensive evaluation process, the plan provides an instrument for control over capital expenditures. The multiyear capital plan should be reviewed and revised annually. The current year should be dropped, the plans for the next several years reviewed and modified as necessary, and the program extended for one additional year to yield a new multiyear plan.

For the capital plan to be useful as a planning and control document, the summary of anticipated capital expenditures should be supported by individual project budgets. Normally, each individual project budget specifies a project number, a description, the purpose and necessity of the project, the project's priority, the scheduled initiation and completion dates, a detailed cost estimate, an estimate of incremental operating revenues and expenses associated with the project, and, if applicable, the estimated return on investment. Although it may be difficult at times to trace the actual source of funds for individual projects, it usually is desirable to identify the intended financing source. This should help identify special financing opportunities, which may reduce the utility's total funding requirement.

Capital-Budgeting Techniques

Several capital-budgeting techniques or measures may be used to evaluate investment alternatives, such as return on investment, internal rate of return, net present value, and life-cycle costing. See financial analysis textbooks for more information on how to use these tools appropriately.

FINANCIAL REQUIREMENTS PLANNING PROCESS

The multiyear capital plan provides a "road map" of capital improvements for a utility. It can guide the utility in carrying out the strategies developed in the system master plan. To make this road map more useful, the utility should also develop a financial plan that describes where the money will come from to construct, operate, and maintain the system after improvement. Many utilities have pushed forward with a capital plan only to discover that they lacked the resources to repay debt or to properly operate and maintain the system in the future. The capital finance planning environment encompasses an iterative process in which a utility must

- determine its current financial condition
- identify capital financing alternatives
- estimate annual capital and operating expenses over a specified planning period
- calculate the amount of revenue required to cover expenses
- assess the rate and service impact on customers
- reevaluate its capital program, if needed, to stay within its financial capability

Current Financial Condition

The first step in developing a financial plan is to characterize the utility's current financial condition. This step will identify the following:

- available financial resources that may be used to finance the capital program, such as existing unrestricted cash reserves, debt authorizations, developer charges, and future revenues
- the existing burden on the utility, including outstanding debt
- the utility's financial capacity to undertake a major capital program

To assess its current financial condition, a utility must have accurate, detailed, and timely data that are generally summarized as information in balance sheet and income statements. Obviously, inaccurate data are not only useless, they are misleading. Even accurate data are of little help if they provide little information to make good decisions. In addition, if the data are out of date, they do little to assist in making timely decisions. Well-managed utilities maintain good financial information systems. Such a system begins with a detailed accounting system that keeps track of revenues, expenses, and investment assets over time and summarizes the results so that management can see how the utility currently stands and what needs to be changed in the future.

A balance sheet, prepared as of a particular date, provides a snapshot of a utility's financial condition. It describes the resources a utility might use to pay for capital improvements, such as available cash reserves, and the liabilities that need to be satisfied, such as outstanding debt. By definition it must balance; i.e., resources (assets) must equal liabilities plus equity. The amount of outstanding debt, what type it is, when it must be repaid, and at what interest rate will all affect a utility's ability to take on new capital projects.

An income statement summarizes the revenues and expenses for a period of time, usually a 12-month fiscal year. It identifies sources of revenue and expenses incurred to produce those revenues. Over the long run, a utility will not be able to take on additional capital projects unless revenues are sufficient to support the projects. The income statement can be examined to assess the amount of cash being generated in excess of current needs.

By using information contained in standard financial reports (e.g., the balance sheet and income statement), a utility can calculate key financial indicators to assess its current financial condition. For example, evaluation of a utility's capital structure—e.g., the ratio of outstanding debt to plant investment—may be one measure of a utility's ability to support additional debt for capital improvements. The financial community typically frowns on having debt funding account for more than 70 percent of total funding.

In addition to financial information, a utility needs information on system use and customer accounting and billing. A fixed-asset accounting system will allow the utility to track by function the original costs, age, and accumulated depreciation of capital facilities; it will assist the utility in deciding whether to repair or replace capital facilities. Comparing annual water bills per household to median household income may indicate whether a utility can reasonably raise rates to support a capital program.

To determine a community's ability to fund a capital program, financial analysts and rating agencies routinely measure financial performance. They rely on a compilation of financial, debt, socioeconomic and management indicators, comparing them to regional and national averages to assess a community's ability to repay debt. By accurately determining its current financial condition, a utility is better able to assess the impact of a proposed capital project.

Other Elements of the Financial Plan

After determining its capital requirements, a utility should identify the financing alternative that will best suit its needs and resources. This process is discussed in chapter 2.

Once a financing alternative is selected, the cash flows that result from it, as well as related operating expenses, must be integrated into a multiyear financial plan and the annual operating and capital budgets. The annual budgets present a picture of where the utility is going over the next fiscal year. They identify what operating and capital expenditures will be incurred, including routine annual capital improvements and capital cost recovery components during the first year of the multiyear capital program.

Revenues by source are typically identified in both the annual budget and the multiyear financial plan. These sources may include user fees, tax revenues, impact fees, and connection fees. As part of this process of identifying revenue sources for the plan, it is often necessary for the utility to evaluate any potential increases in or modifications to its revenue sources that may be necessary to support the identified capital and operating expense estimates. If additional revenues will be needed, the utility must estimate how its costs would be affected by increases and whether the increases would impose an undue financial burden on system users. If the increases associated with a particular alternative are perceived to be too significant, the planning process must be fine-tuned to reduce project cost, to identify alternative financing, or to find different sources of revenue to support the financing.

AWWA Manuals M1, *Water Rates*; M26, *Water Rates and Related Charges;* and M34, *Alternative Rates,* strongly support the association's Statement of Policy regarding financing and rates, which stresses that the public is best served by revenue enterprises that are adequately financed through rates based on sound engineering and economic principles and that do not require subsidies from other entities. However, sensitivity to financing and rate impacts on customers has come to the forefront in the development of optimal financial plans. Table 1-1 provides a checklist of questions about the impacts on system stakeholders that should be addressed during the development of an optimal capital financing plan.

Table 1-1 Financial plan and rates checklist

Affordability Check

Yes	*No*	
☐	☐	Are there unwarranted negative impacts on specific customer groups?
☐	☐	Are there unintended negative impacts on the local economy?
☐	☐	Has the public expressed concerns over its ability to pay?

Implementation Practicality Check

Yes	*No*	
☐	☐	Are there limitations of metering devices?
☐	☐	Will the rate structure switch affect customers fairly?
☐	☐	Can utility billing procedures and software adapt to a change in billing frequency?
☐	☐	Will customers accept large periodic rate increases instead of small annual rate increases?
☐	☐	Are variations in rate impacts by customer class acceptable to the public?
☐	☐	Are the cost and schedule of implementation reasonable?
☐	☐	Does the utility have acceptable staff levels?
☐	☐	Is there enough expertise on staff?

Financial Forecasting

Strategic financial planning requires projections of utility revenues and costs over a planning period that can range from 3 to preferably 5 years for a capital improvement plan to as many as 20 years for a strategic master plan. Although projections for periods longer than 5 years become increasingly speculative, they provide insight into long-term impacts of today's business decisions. Economic and regulatory conditions affecting future service requirements are uncertain, but long-term capital planning requires that the effects of the capital budget on revenues and revenue requirements be demonstrated for a specified planning period.

Forecasts show proposed changes (1) in capital investment as a result of anticipated changes in service requirements, (2) in costs because of inflation or operating trends, and (3) in revenue because of growth. Generally, these projections are based on historical trends adjusted for any known changes in future conditions. Trends in such indexes as the consumer price index and construction cost indexes may be used to adjust current costs into the future.

Projections over the planning period can generally start with the annual budget. Revenues, operations and maintenance costs, and capital expenditure projections are extrapolated over the planning period and adjusted for inflation, growth, and other anticipated changed conditions. Specific project-related operations and maintenance costs should also be included, e.g., additional operating personnel to staff a planned new water treatment plant.

Financial Planning: An Iterative Process

A typical complaint about financial planning is that, immediately upon completion of plan development, the first exception to the plan that arises makes the plan obsolete. Such is the reality of planning in general. This reinforces the need for financial planning to become an iterative process in which utility management engages in continual stages of review, projection, evaluation, revision, and implementation so that any needs for additional income or funding sources to meet the utility's revenue requirements are promptly identified. The process results in a living document when the plan is updated at least annually.

As discussed earlier in this chapter, broad evaluations of the financial impact of estimated future conditions should be based on multiyear analyses. Management should always consider the long-term impacts of bonding capacity, rate levels, and affordability for individual customers when planning major projects. Significant risk exists in planning for only a 1- or 2-year horizon. Evolution of current-year planning into long-term perspective is imperative to minimize the possibility of making poor business decisions today that have long-term ramifications. Utility management needs to know "what's over the hill" in order to make sound financial decisions today.

SUMMARY

The process for developing an integrated and cost-effective capital program and financial plan should include the steps summarized in Table 1-2.

Table 1-2 Developing an integrated and cost-effective capital program

Step	Actions
1	Develop formal CIP documentation for a 5- to 10-year period. Try to avoid nearsightedness in the planning process.
2	Prioritize all projects based on established criteria that fit the community's growth patterns and customer expectations for service.
3	Seek advice from stakeholders and keep them informed about the utility's operational and financial needs and proposed rate adjustments. Consider the use of formal public hearings or advisory committees or both to fully explain utility capital needs and financing requirements and to engender public support before rate adjustments. Be sensitive to customer budget or planning cycles through proper notification of rate revisions.
4	Understand the iterative process of capital and financial planning and rate level adjustment determinations. Evaluate the integration of funding sources, bonding requirements, financial risks, and impact on rates—either individually or in total—before identifying or ranking projects for the capital program. Consider ways of controlling the impact on rates by shifting risks to growth customers through the use of system development charges and other developer-related charges.
5	Integrate performance measurements into a utility's capital program and financial planning to assist stakeholders in assessing accountability and in making economic, social, and political decisions.

Chapter 2

Identifying Financing Alternatives

Once a utility's financial plan has been updated for the current year, it is time to explore financing alternatives. The goal is to identify the alternative or combination of alternatives that provides the utility with funds needed to support the identified capital program in the most effective manner. In other words, how can capital improvements be paid for without causing rate shock to the utility's customers?

Possible alternatives for financing capital improvements include internal sources, such as excess cash from operating revenues, as well as external sources, such as debt, equity, grants, and leases. Privatizing or outsourcing of some operations may also be a substitute for some types of capital investment. This chapter defines and discusses the various types of funds available to water utilities.

After the various financing alternatives have been identified, the utility should explicitly evaluate the advantages and disadvantages of each source of funds. The evaluation process is discussed in chapter 3.

INTERNAL FUNDING

Many utilities pay for some system improvements with internal funding. This pay-as-you-go financing is increasingly being used to reduce the interest burden that accompanies debt financing; it is often particularly appropriate for short-term and lower-cost capital projects. Internal sources can include rates or operating income, impact fees, system development charges (SDCs), and contributions. To maximize the portion of operating revenues available for capital financing, utilities often look at one or more of the following: (1) cost reductions or cost avoidance, (2) revenue enhancements, and (3) including depreciation as a cost to be recovered through water rates.

Cost Reductions and Cost Avoidance

Within the category of cost reductions and cost avoidance, many actions can increase the cash available for capital projects. Among the actions available to most utilities are:

- process improvements
- energy cost savings
- outsourcing and contract operations

Process improvements. An operational review to maximize the efficiency of an organization can often locate funds for capital improvements. Process improvement involves breaking down operations into processes and then evaluating the effectiveness of each process. Once each process has been reviewed, operations can often be streamlined by eliminating wasteful practices, introducing standard practices, providing better training, or using automation. Opportunities for savings, such as through competitive bidding or the use of fixed-cost annual contracts, are often found as a result of an operational review.

Why is this an effective cost-reduction method? Processes evolve over time and often contain duplicate tasks or outdated methods. This adds time to processes, and additional time usually results in increased labor costs or duplicating work. A high number of approvals required in the course of a process often indicates that an operations review is needed. Approvals are built into a process that at some earlier point did not consistently produce the desired results; the approvals were added to the process to catch the errors. By eliminating each unnecessary step and improving the remaining steps, a utility can improve the process itself, thereby spending less money on materials, equipment, and labor. These savings can be used to pay for capital improvements.

Energy cost savings. Energy to operate treatment plants and pump water throughout a distribution system represents a significant operating expense for water utilities. Rate structures employed by electric utilities often include substantial costs for high energy use during peak usage periods. This poses a problem for utilities because customer demands for water service and electric service often coincide. This leads to the very result that the utility wants to avoid: high electric usage during peak periods. However, there are solutions. For example, gas-driven pumps can be installed and used during peak electric use periods to prevent the utility from being placed in a higher rate bracket. Alternatively, water towers can be filled during off-peak hours so that additional pumps do not have to be turned on during the peak hours.

Electric and gas companies, as a rule, have multiple rate structures. In most cases, a utility could potentially meet the criteria for any of several structures. By becoming familiar with the structures and negotiating for the most advantageous one, a water utility can realize significant savings. Experts with experience negotiating advantageous rates with power companies can assist water utilities in this area.

In addition, some energy utilities have different rate incentives for different types of fuel. The water utility should account for these rate differentials in the analysis of what types of pumps to use. Converting from one fuel source to another may also at times be economical because the energy utility may be offering incentives that cover some or all of the conversion costs.

Outsourcing and contract operations. Water utility operations include a wide variety of activities. While most of these activities can be performed more cost-effectively by the utility—because the utility does not need to make a profit on the activity—some activities are more efficient when performed by others. These are the activities that should be considered for outsourcing. For example, sometimes significant capital outlays are required for equipment that is only seldom needed. Private businesses may be able to buy this equipment and contract with several entities to use it more fully. Because the cost of the equipment is spread over several entities, everyone can save money.

Contracting can also be used to handle peak activity that base staffing levels cannot support. For example, an engineering division can be staffed at a level to handle normal design activity, and consultants can be employed when activity levels exceed the capabilities of the staff.

A useful exercise when evaluating a utility's efficiency in performing an activity is to determine the cost of outsourcing that activity and then require the utility to meet or beat that cost. If the utility cannot operate at a cost that is competitive with outside entities, the utility should either re-engineer the process in order to become competitive or contract the function. Some utilities even have internal groups formally bid against outside contractors as a way to maintain cost-sensitive pressure on the utility.

Revenue Enhancements

Most utilities receive revenue from several different types of customers, such as

- developers (to support system expansions)
- builders (to offset the cost of installing new meters)
- water customers (to pay for the cost of treating and delivering water)

Utilities should establish fees and charges that recover the cost of providing service to each group in order to prevent unintentional subsidizing of one group by another. Also, efforts should be made to ensure that all service provided is billed. For example, programs should be established that minimize unaccounted-for water. Such water is costly because money is spent to obtain and treat water that is never sold. Revenue can be increased or costs reduced through testing and changing out faulty meters or through a leak detection survey to identify needed repairs. More extensive information on water meters can be found in AWWA Manual M6, *Water Meters—Selection, Installation, Testing, and Maintenance.*

Depreciation

Including depreciation as a cost to be recovered through water rates is a method of generating revenues internally to fund capital improvements. The selection of a depreciation method (such as original cost versus replacement cost) to be included in the rate structure will affect funds available in any year. Depreciation approximates the value of the infrastructure that needs to be replaced over time. If appropriate levels of depreciation are used to set rates and those internally generated funds are used for the replacement of worn-out capital infrastructure, a utility will be in a position to maintain financial integrity and limit the use of debt financing.

EXTERNAL FUNDING

As discussed in the previous section, internal funding mechanisms frequently finance some of a utility's capital needs and are an appropriate method for short-duration, low-cost projects. Utilities often pay for such projects on a pay-as-you-go basis, i.e., from existing customers through rates for service. For more costly capital projects, however, utilities usually need to find an outside source of money. External sources can include debt, equity contributions (such as grants), customer contributions, or private sector investment. To stay financially healthy, a utility should balance the manner in which it finances capital projects with a combination of internal and external funding. This section describes each of these external funding sources and their characteristics. Chapter 3 discusses some of the advantages and disadvantages of each source and provides suggestions for evaluating each source of funding.

Debt Financing

Major capital projects are frequently funded with some form of debt. Debt financing allows a utility to raise more funds for capital projects than would be feasible on a pay-as-you-go basis, and it allows the utility to repay those funds over time. Debt can be repayable over various periods. The length of time between borrowing and fully repaying determines whether the debt is considered long-term or short-term debt. Debt that must be repaid in 1 year or less is usually considered short-term debt. Debt that is repayable over longer periods may be called intermediate- or long-term debt.

The nature of the promise to repay debt may differ in various ways. Some debt may require that a portion be repaid each period; other debt may have periodic interest payments but require the full principal amount to be repaid at the end of the term. The utility may make a general promise to repay debt, or it may provide specific collateral as security for its repayment obligation.

Debt may be raised from various sources, both public and private. How easily a utility can access these sources and how much it will be able to borrow will depend on its size, its ability to repay the debt, and the impact the debt will have on its rates.

Commonly Used Forms of Short-Term Debt

Short-term debt may take the form of bank loans, notes, commercial paper, or floating-rate demand notes.

Bank loans. Loans from banks can be structured as term loans or as lines of credit. A term loan is repayable at a specific future date. A line of credit is an arrangement under which the borrower can borrow and repay money from time to time as needed. A line of credit may be committed or uncommitted. Under an uncommitted line, loans are made at the bank's discretion with terms set at the time of borrowing. In contrast, with a committed line of credit the utility and the bank enter into an agreement that (1) sets the terms under which borrowings may be made and (2) obligates the bank to honor requests for borrowings during the duration of the agreement as long as the borrower meets predetermined credit criteria. A committed line of credit generally requires that the borrower pay a fee to the bank in return for having a source of funds available.

Bond-, tax-, grant-, and revenue-anticipation notes. A note is an unsecured debt, usually with a maturity under 10 years. The various types of anticipation notes are promises by a municipal issuer to repay the borrowed amount from a source of money that the utility will receive in the future, such as a future bond issue (in which case the note is called a bond anticipation note, or BAN), anticipated revenues (revenue anticipation note, or RAN), taxes (tax anticipation note, or TAN), or grants (grant anticipation note, or GAN).

Commercial paper and tax-exempt commercial paper. Commercial paper (CP) is an unsecured short-term promissory note with an average maturity of 30 to 45 days. The notes, which under security regulations have a maximum maturity of 270 days, are intended to be refinanced (rolled over) continuously for periods that may exceed 1 year. CP may be issued by investor-owned utilities or by publicly owned utilities. When CP is issued by a state or local government, it is called tax-exempt commercial paper (TECP). CP or TECP differs from other interim financing in that it is generally rated by a national rating agency and is supported by a backup source of funds, such as a bank line of credit, a letter of credit (LOC), or another credit facility, in order to protect the purchaser of the notes from default in the event the utility cannot sell a new issue when a maturing issue becomes due. The use of TECP by municipal issuers has increased measurably in recent years because TECP is a prime investment instrument for tax-exempt money market funds.

Floating-rate demand notes. Floating-rate demand notes are securities on which the interest rate changes at predetermined intervals (often monthly) and that give the purchaser of the security the right to demand that the seller (utility) redeem the notes at each interest adjustment date. Floating-rate notes are usually issued in $100,000 denominations and are placed primarily with money market funds or other large institutional buyers.

Commonly Used Forms of Long-Term Debt

Long-term debt may take the form of a note, a debenture, or a bond. Both notes and debentures are unsecured borrowings. Notes generally have maturities under 10 years and debentures have maturities of 10 years or more. While the term *bond* in strict usage means a secured debt, in general usage it can refer to various kinds of secured or unsecured debt. In this manual, it is used to refer to all types of long-term debt.

General obligation bonds. General obligation bonds, referred to as GO bonds, are debt obligations issued by a government entity, such as a state or local government, and backed by a pledge of the entity's full faith and credit. GO bonds can be issued only by units of government that have the authority to tax. Water utilities that are separate from local governments may not have this authority. GO bonds are secured by an unconditional pledge that the issuing government will, if necessary, levy unlimited taxes to repay the debt. This is the strongest pledge the government can provide, and it is usually regarded by investors and rating agencies as the strongest form of bond security. Therefore, a GO bond generally has the lowest cost of the various financing vehicles available to local governments. Because GO debt is issued by a state or local government, interest paid to the lender is exempt from federal income taxes. Depending on state tax laws, this debt may or may not be exempt from state income taxes.

Revenue bonds. Revenue bonds are debt obligations issued by local governments or other public agencies. Principal and interest on revenue bonds are secured by (and may be payable only from) the specific revenues named in the bond documents. Revenue bonds are commonly used to finance activities that generate revenue and that are expected to be self-supporting, such as a water utility. Debt service payments on revenue bonds issued by water utilities are usually made from charges paid by users of the water system.

With a revenue bond, debt repayment may depend on timely completion of a project, on adequate rate or charge structure, and on sound fiscal management of the enterprise. For this reason the bonds are assumed to be riskier and will have a higher interest rate than a GO bond issued by a comparable issuer. Investors' acceptance of revenue bonds is highly dependent on the service or project to be financed and on the creditworthiness of the borrower.

Most revenue bond agreements or indentures contain provisions, or covenants, that are intended to enhance marketability by providing investors with additional assurances that their money will be repaid. The following are among the provisions typically included in revenue bond documents:

- A rate covenant, by which the issuer agrees to set rates sufficient to meet all operating costs and some multiple of debt service. For example, there may be a requirement that net revenues available after operating costs will be sufficient to cover 125 percent of debt service.
- A flow-of-funds requirement that specifies the order in which revenues will be used. For example, there may be a requirement that revenues be used first to pay operating costs, then for debt service, then to replenish reserve funds, and then for costs associated with system repairs or replacement.
- The establishment of a fund to set aside money to ensure debt service payments are made on a timely basis or to cover debt service for some period of time in the event of a revenue shortage. This deposit of money is called a debt service reserve fund, and the indenture generally requires that the fund be administered by a third party, such as the trust department of a bank. For example, many agreements require the amount to be equal to the maximum debt service in any year during which the bonds are outstanding.
- An additional bonds test, which provides that additional bonds may not be issued unless historical and projected revenues indicate that there is sufficient revenue to avoid dilution of coverage on outstanding bonds.

Sometimes a government adds a GO pledge to a revenue bond backed by a specific source of revenues, such as the net revenues of a water system. These bonds are sometimes referred to as double-barrel bonds because they provide the dual security of a dedicated revenue source with the general taxing power of the government.

Bonds with credit enhancements. A small municipality or utility may have difficulty issuing debt, especially if it has not issued debt before, does not have a credit rating, and is unknown to the investment community. One option available to this type of organization is to borrow from a local bank or other institution where the organization's creditworthiness is more easily demonstrated.

Another option is to obtain a credit enhancement, generally in the form of bond insurance or a letter of credit from a more well-respected entity. Under such an arrangement, the utility will pay a fee to have the better-known party, such as an insurance company or bank, guarantee to the investor that it will ensure the timely payment of principal or interest or both. The addition of a guarantee by a large, well-known entity with high credit ratings makes a bond more attractive to investors; therefore, the interest rate on the bond will be lower. Unless the savings in interest cost over the life of the bond is greater than the fee—thereby making the total cost of debt lower to the issuer than the cost of trying to issue debt without third-party assistance—credit enhancement is not recommended.

Governmental loans. Federal and state agencies provide a valuable source of lower-cost financing through a number of programs that supply infrastructure financing. The federal Rural Development Agency (formerly the Rural Development Administration) provides loans to rural and economically depressed areas for water and wastewater system improvements. Amendments in 1996 to the Safe Drinking Water Act establish a State Revolving Fund (SRF) program to provide loans to water utilities for capital improvements at interest rates at or less than market rates. A similar program under the Clean Water Act has supplied low-rate loans for wastewater treatment facilities for a number of years. The financial advisor should be aware of current sources of governmental loans and should be consulted about the structure and terms of loans available to the entity.

Leasing

A lease is an arrangement between the owner of property (lessor) and the user of property (lessee) that gives the lessee the right to use the property for a defined period in return for meeting certain requirements, including the payment of rent. Leases may be attractive for municipal utilities as a means of acquiring the use of needed equipment and facilities when debt limitations restrict direct purchase and ownership. Leasing in some of its forms is treated as debt for financial reporting purposes, and leasing is generally viewed as a substitute for debt financing.

Because leasing often involves tax benefits to the lessor and tax implications to the lessee, and because it is subject to the interpretations of the IRS, it is a complex form of debt financing. Tax acts passed in 1984 and 1986 substantially reduced the benefit of tax-oriented leasing. Utility managers should retain experienced legal assistance when considering a lease because tax laws are constantly changing.

Sources of lease financing include large banks, equipment manufacturers, real estate development firms, and major leasing companies. Most firms provide preliminary lease-rate quotes, which may be used to compare lease costs against ownership costs. The major categories of leasing likely to be encountered by a finance staff that is considering how to finance a new facility or new equipment include direct leases, leveraged leases, and certificates of participation.

Direct leases. With a direct lease, the lessee negotiates the equipment or facility specifications, as well as the rental terms, with the lessor. After the lease is signed, the lessor purchases the equipment or constructs the facility exactly as specified in the lease agreement. When the equipment or completed facility is delivered, the lessee makes sure all specifications have been met, then formally accepts the facility.

Leveraged leases. A leveraged lease is similar to a direct lease (which is sometimes called a nonleveraged lease), but it is more complex because another party is involved. The additional party is the entity that provides funds to the lessor for constructing the facility or acquiring the equipment.

Certificates of participation. A certificate of participation (COP) is a security backed by an interest in a stream of rental payments. The lessor is generally a governmental entity, often one specifically created to enter into lease transactions, and payments received by certificate holders are generally tax-exempt. The financial advisor should be able to explain the current treatment of a COP transaction under any local debt statutes; the advisor should also be able to explain the current market requirements with respect to common restrictions concerning use of property, the flow of funds, security for the certificate holder, and the basis for tax exemption. Restrictions placed on the lessor concerning use of property and disposition of lease payments establish security for the certificate holder and the basis for tax-exempt interest payments from local government.

Contributions

Customer contributions of money or property are a common and important form of capital for water utilities. Historically, customer contributions have been related primarily to customer connections and distribution system facilities. Customers are frequently required to purchase their meters and pay for the installation of taps and services. Developers are often required to provide distribution mains, fire hydrants, and customer connections for new subdivisions as contributions in aid of construction. While these contributions relieve a water utility of some of its capital needs, they often do not provide financing for supply, treatment, or distribution facilities.

Some utilities require developers to make customer advances for construction of subdivision facilities. The utility may then reimburse the developers during a specified period as new customers connect to the system. Any funds remaining at the end of the period are retained by the utility as contributions in aid of construction.

Publicly owned utilities may also include funds provided through intergovernmental transfers as contributions if those funds are not intended to be repaid from utility revenues. This type of contribution is often used to cover a portion of capital requirements for specific facilities, and the funds may be provided by the local government from GO bond proceeds, property, or sales tax revenue.

Grants

Development of public water supplies has traditionally been a local responsibility. Unlike the wastewater industry, the drinking water industry has not had a large federally funded grant program. With reduced federal spending and a trend toward favoring loan programs, little if any federal grant money is expected to be available to water utilities in the near future. This is true even as federally mandated drinking water quality standards require additional capital investment by local utilities. As of the time this manual was written, the trend appears to be toward reducing the number of grants available even to wastewater entities. For example, the construction grants program funded under the Clean Water Act for wastewater utilities is being replaced by a state revolving loan program.

In spite of this trend, there are a few grants available from various federal and state agencies for rural or economically depressed areas. The federal Rural Utilities Service provides grants and loans for rural water and sewer systems and communities with populations of fewer than 10,000.

SUMMARY

There are many sources of funds for capital improvement. Among these are internal sources, such as savings from improved operations and revenue enhancement, and external sources, such as grants and loans. Because of the size of most CIPs, utilities generally rely on debt for a large proportion of their capital-financing needs.

Chapter 3

Evaluating Financing Alternatives

Whether a particular funding alternative is the best choice for a particular project is influenced by the legal environment, financial factors that affect all issuers, financial considerations specific to the issuer, and other external factors. Legal factors may restrict the size or type, or both, of debt that a utility may use. Financial factors include the condition of the capital markets, as well as the effect the alternative will have on the utility's future financial flexibility. External factors include whether the alternative is likely to be accepted by governing authorities and the public. This chapter discusses these factors. Some advantages and disadvantages of the various financing alternatives discussed in chapter 2 are also discussed in the final section of this chapter.

LEGAL FACTORS

The first step in evaluating a proposed financing alternative is to determine whether that alternative is available to the utility under the various laws, regulations, and governing documents to which the utility is subject. The financial advisor and bond counsel, along with in-house counsel, should be able to discuss the various laws and regulations that are relevant to the availability of a particular financing alternative. If the alternative is not currently permitted, the utility may want to evaluate the costs, time requirements, probability of success, and risks involved in seeking a change in its operating environment. These factors can then be evaluated against the benefit that the utility would gain if the alternative were available. If the changes needed to make the alternative available are not feasible, the utility should seek another option.

An investor-owned utility may be subject to rules promulgated by the local regulatory body, as well as to previous indenture language, corporate resolutions, and language in the utility's by-laws. A publicly owned water utility is often subject to restrictions in its governing documents, as well as local and state laws. Both types of utilities are subject to various federal laws if they wish to borrow on a tax-exempt basis.

Governing Documents

Legal factors affecting debt issuance are frequently found in a utility's own enabling documents, whether those documents represent the original authorization or subsequent actions of the utility's board or other oversight agency. The governing documents for a publicly owned utility frequently establish limits on borrowing, either as an absolute cap or as a ratio. They may also identify what types of debt can be issued, including negotiated or competitive sales, restrictions on private placements, selection of the finance team, and other considerations. Most jurisdictions also have established rules about the investment of bond proceeds, with an emphasis on the preservation of capital.

Additional legal restrictions may be established by legislation or board action authorizing previous debt issuance. For example, a previous resolution may have already pledged certain assets as security for another debt issue. As a result, assets and revenues may not be available to support an alternative currently under consideration unless the new debt is subordinated to the old. This may affect the feasibility and cost of the alternative.

State Law

State laws frequently address the legal structure and borrowing capacity of local government entities, including municipal utilities, and utility authorities. Where requirements exist for voter approval of debt issuance, they are generally found in state statutes.

Federal Law

Federal law affects utility borrowers in several ways. The Securities and Exchange Commission (SEC) has extensive regulations that apply to public debt offerings made by investor-owned utilities. In addition, through its regulation of the underwriters, the SEC also mandates that publicly owned issuers make certain information available regarding the utility's financial situation beginning at the time of borrowing and continuing until the debt is retired. Although the issuer's bond counsel often handles many of these activities (see "Bond Counsel," page 46–47), the issuer has the ultimate responsibility for ensuring compliance. If the utility, whether investor owned or publicly owned, wishes to borrow on a tax-exempt basis, it also needs to comply with arbitrage regulations concerning (1) the maximum amount of interest that may be earned on borrowed funds after the transaction is completed and before the funds are expended on capital projects and (2) how long the funds can be held before the issuer is subject to penalties and possible loss of the tax-exempt status of the issue.

Bond counsel and the financial advisor should be able to explain what is required from the borrowing utility in such a way that the utility can assess the cost or administrative burden or both associated with the various laws and regulations when evaluating the appropriateness of a financing alternative. If the utility is for some reason unable to comply with these requirements, it should choose other financing alternatives.

FINANCIAL FACTORS

Before making any decision regarding the way a capital project will be financed, a utility should carefully consider financial factors that influence which financing plan is most appropriate. If the plan includes new borrowing, financial considerations are also a factor in deciding which debt instrument is most appropriate. Among the financial factors specific to the issuer that need to be considered are (1) whether the proposed alternative is compatible with the overall objective of the financing program and (2) whether selecting the proposed alternative will increase or reduce the utility's flexibility in the future. If, as a part of its long-term financing objectives, the utility has set targets for such things as the proportion of its debt that will be variable rate, the maximum amount of debt it will incur as a percentage of fixed assets, or a maximum annual amount for debt service per connection, it should evaluate how the proposed alternative will help or impede it in meeting these targets. Financial factors that affect all borrowers include whether there is a general consensus about the level and direction of interest rates, the outlook for economic activity, and whether there are proposed legal or tax changes that make borrowing more or less difficult than usual. Several of these factors are discussed in more detail in the following paragraphs.

Ability to Meet Covenants and Satisfy Revenue Requirements

The indenture requirements of existing debt are similar to the legal factors discussed previously in that they may contain provisions that affect a utility's ability to issue new debt. For example, existing bond covenants may specify that rates be established at a sufficient level to meet certain requirements. The documents may also require the utility to meet specific coverage ratios (which may be calculated on either an accrual or a cash basis). The utility must be able either to satisfy the covenant or to amend the provisions when the new debt is issued. An amendment may not be possible unless the old debt is repaid.

Even if the plan is to repay the old debt, investors may not be willing to accept a lower coverage ratio without a significant increase in the interest rate unless the revenue stream is very predictable. A lower coverage ratio may also mean the bonds will receive a lower rating. In evaluating what coverage is appropriate, investors look at historical data in conjunction with projections, and they assess the reasonableness of the assumptions on which projections are based. They do this in order to evaluate the quality of the revenue stream pledged to pay bondholders. They will examine intraperiod fluctuations and their causes, as well as the sensitivity of the revenue stream to adverse external events. The more volatile the revenue stream (and the greater the risk), the more coverage is needed to attain as much protection as an issue with lower coverage but higher-quality revenues. The issuer should perform a similar analysis (1) to assess the reasonableness of being able to obtain a lower coverage ratio and (2) to be able to answer questions that arise during the marketing process.

In using the financing plan to assess the impact of new financing, the utility should make sure that both its revenue and its cash needs are met over the planning horizon. This means that sufficient working cash must be available after debt service on the new debt is considered. As noted in chapter 1, this is often an iterative process that balances financial requirements with financing options. As an example, if utility rates are to be left unchanged for 2 or more years, the financial plan should show that sufficient reserves, such as a rate stabilization fund, will be built up during the first half of the rate period to finance an expected deficit in the latter half of the rate period. If sufficient revenues are not shown, the planner may want to evaluate (1) whether the utility's governing board would be willing to consider annual, biennial, or triennial rate increases, or (2) whether it is possible to restructure the new obligation in a way that fits better with existing obligations in order to enable the utility to meet both covenants and cash requirements during each rate period.

Impact of Debt Portfolio

If debt is an alternative that is being considered, the issuer needs to be aware of the amounts and types of debt currently outstanding and how those obligations will influence or be influenced by the new financing. Some specific concerns regarding the existing debt portfolio include

- The structure of existing debt relative to new debt. How will the repayment terms of the new debt fit in with existing obligations? Are the same revenues pledged? Are payment and reporting dates the same? What will this mean in terms of administration? If there is variable-rate debt, when will rate changes occur on the various obligations, and what will this do to the organization's overall risk profile and budgetary process?
- The need and ability to create relative levels of subordination for debt issued at different times for different purposes. Can new obligations be secured equally with old obligations? If not, how will that affect the marketability of the new debt?
- The ability and desire to refund, or defease, existing debt to do one or both of the following: take advantage of improved interest rates and facilitate revisions to bond covenants. Are there restrictions in current indentures that could be solved with a refinancing? Have interest rates moved sufficiently so that refunding is economically justified? How will the limitations on advance refunding of tax-exempt debt affect what alternatives are available? Investor-owned utilities will also want to be aware of how the transaction will affect reported earnings per share and what effect this may have on a given utility's stock price.
- The impact of current indebtedness on the borrower's creditworthiness. Is the utility more or less leveraged than its peers? What effect will another debt issue have on the credit rating? How will that debt issue affect the prices of currently outstanding debt? What effect will an adverse impact on the price of outstanding debt have on the marketability of the utility's new debt (and, for investor-owned utilities, on stock price)? Does the utility have the ability to raise revenue, in the case of revenue bonds, and to meet the provisions of the rate covenant? What effect will the ability or inability to raise rates have on the utility's credit rating and ability to increase the amount of debt outstanding?

Market Conditions and Access

Utilities that issue new bonds or refund existing debt should be aware of those market factors that influence financial decisions. The bond markets have tremendous diversity. Corporate markets as well as the municipal markets are influenced by the economic outlook, US Treasury activities, and general investor sentiment. In addition, the municipal markets are influenced by proposed federal tax law changes, as well as by proposed changes in the requirements and specifications for issuance that are promulgated by the states.

If the markets are in an unsettled state, there could be advantages to using a negotiated sale or private placement (if available) rather than a public competitive sale because the issuer would have more flexibility in settling the time of sale. To properly assess the impacts of these market-related concerns, issuers typically rely on competent professionals in a variety of fields, including bond counsel, the financial advisor, and the underwriter. The assistance provided by these professionals is discussed in chapter 5.

EXTERNAL FACTORS

After a utility has determined that it is legally able to pursue the financing option it has selected and (if it has selected a borrowing) has assured itself that it will have the resources to repay the loan on a timely basis, it must address certain external considerations. Among these are whether the utility has public support for the capital improvements and for the financing plan to be used. These two matters are usually intertwined, and often the financial factors drive public debate on the capital program.

Most borrowings by publicly owned utilities require citizen approval, either directly through referendum or indirectly through actions of an appointed board or elected council. Investor-owned utilities may need board approval, regulatory body approval, or both. Accordingly, the capital-financing vehicle should be selected with consideration for its acceptability to the general public, the utility's authorizing body, or both. The utility staff should consider its ability to explain, justify, and successfully advocate the capital items to be financed, as well as the financial assumptions underlying the borrowing. A significant factor influencing public acceptance is the impact on customer costs. In jurisdictions where a ballot initiative is required for debt issuance, utilities should take particular care to evaluate the political and financial climate before approaching voters.

Securing necessary administrative and legislative approval should be taken no less lightly. The need for public approval, whether directly or through elected or appointed representatives, should not deter a utility from using a financing mechanism it believes is appropriate and well suited to a necessary and specific capital program. Typically, a review and reporting of operations and budgets, as part of continuing disclosure, provides an added level of security. However, borrowers should recognize that more complex and innovative financial products will face closer scrutiny.

The level of scrutiny for publicly owned utilities has increased recently because the federal Securities and Exchange Commission has clarified the responsibilities of elected and appointed officials to understand the structure and risks inherent in bond issues to which they give their approval. As a result, even traditional borrowing vehicles now sometimes face questions from overseers who are not familiar with such arrangements. Issuers who contemplate using derivative products and innovative or nontraditional financing should plan carefully to explain their proposed transaction to boards or elected bodies.

ADVANTAGES AND DISADVANTAGES OF COMMONLY USED FUNDING ALTERNATIVES

This section describes some advantages and disadvantages of funding alternatives commonly used by water utilities, including general obligation bonds, revenue bonds, and structure variations. Definitions and characteristics of these options are given in chapter 2.

Grants

Using grant funds can generally help utilities keep total costs of new construction low for current users. Entirely grant-based funding of capital projects does not necessarily lead to an increased level of required utility revenues to support operation of the new facilities. However, with partial grant funding there may be a tendency for local governments to oversize systems or facility capacity because the utility's money goes considerably further. Because such oversized facilities may not work as efficiently, oversizing may result in higher operating expenditures and greater annual cash expenses. This will impact user charges for the life of the facilities.

Most grant funding requires increased reporting and strict compliance with grant provisions. This could be a disadvantage if the utility's needs change in the future, since a change in use may require the utility to incur unforeseen expenditures at a future date or require outright repayment of the grant itself.

Contributions

Contributed funds permit a utility to ensure that new users share in the full cost of developing utility service capability. Contributions may be used to expand service into new areas or reduce effects of rehabilitating or upgrading facilities. In either case, the utility may keep the level of costs that need to be recovered from current revenues lower than might otherwise be the case. Contributions may help postpone or reduce the size of future rate increases. They may also pay for reserve capacity for which existing customers may otherwise have paid the carrying costs over a number of years. Through a utility's use of contributed funds, current customers are less likely to be asked to subsidize future growth or rehabilitation.

A disadvantage of contributed funds is their inconsistent nature. Receipts usually depend on growth cycles and the demand for utility services. For those capital projects in which a utility shares construction costs with others, the utility may have particular difficulty planning the timing of such project expenditures.

Leases

Leases often provide additional flexibility to a utility. For example, a leased facility can often be obtained more quickly than one constructed by the utility.

Leases can be offered at taxable or tax-exempt rates. A lease at a taxable rate may be more expensive to a publicly owned utility than using funds from other sources. However, if the utility has only a short-term need for specialized equipment, or if equipment is expected to become obsolete before it wears out, the utility may find it advantageous to lease the equipment, even at the higher rate, especially if the equipment can be returned or upgraded before reaching the end of its useful life.

The most obvious benefit of a leveraged lease is the low cost. Lessors can pass on a portion of their tax savings to the lessees, resulting in a lower cost for the latter. A disadvantage of a leveraged lease is the complexity and initial cost in arranging such financing. Transaction costs of 8 to 9 percent may be involved.

A lease in the form of certificates of participation has many of the same advantages and disadvantages as a revenue bond. (Revenue bonds are discussed later.) The cost to the utility to obtain funds through the use of a COP structure depends both on the credit rating of the utility and on how essential the leased facility is to the utility. A COP secured by essential facilities is more attractive to investors because it is generally secured only by rental payments on the leased facility.

General Obligation Bonds

General obligation bonds are usually the lowest-cost form of debt financing available to local government. In addition, investor familiarity with this traditional standardized form of financing makes GO bonds very marketable. Costs of issuance are generally lower than those associated with revenue bond financing, not only because the cost of marketing a bond with wide market acceptance is lower, but also because the documentation associated with a GO issuance is less complex than that required for revenue bond financing. Because GO bond financing does not require the restrictive covenants, special reserves, and higher debt service coverage associated with revenue bonds, current revenues are not overly restricted, or pledged, and can be used for other purposes.

While the issuance of GO bonds has many advantages, there are also certain drawbacks to this form of financing. One of the most significant disadvantages is that increases in the amount of GO debt must generally be authorized by the electorate. At a minimum, this increases the time needed to finance. In addition, many states limit the amount of GO debt that can be issued or the interest rates that can be paid. These debt ceilings or interest rate ceilings on GO bonds may be restrictive in areas where tax reform propositions have been passed. Additional local government restrictions may exist in regard to the length of amortization of the bonds; this would tend to make the GO bonds' structure somewhat less flexible than with other forms of borrowing.

Revenue Bonds

Revenue bonds offer important benefits to the issuer. They provide a mechanism for recovering costs of specific services, such as water supply, directly from the users that benefit from the services. Revenue bonds are available to agencies that do not have taxing authority to issue GO bonds. Unlike most tax-supported debt, revenue bond issuance does not usually require voter approval; although voters generally have a right to petition for a referendum. Revenue bonds also avoid possible dilution of the government's pledge of full faith and credit, since such bonds may not be included in the issuer's debt limit.

Revenue bond financing also has some disadvantages. Because principal and interest on revenue bonds are secured by and payable exclusively from revenue received from system operation or from the project being financed, market acceptance of these bonds is highly dependent on the service or project to be financed and on the nature and reputation of the issuer. For this reason, revenue bonds generally command higher interest rates than do GO bonds.

Higher interest rates and marketing difficulties are particularly apparent with regard to enterprises that do not have established earnings potential or a history of financial operations. Revenue bonds also frequently contain numerous provisions that make a revenue bond borrowing more administratively cumbersome than a GO borrowing. Indenture provisions—such as requirements for special reserve funds and debt service coverage ratios, as well as limitations regarding the amount of future debt that can be issued—may also decrease the issuer's future flexibility. The provisions often make it difficult to issue additional bonds, and they may trigger refinancing or refunding of outstanding debt before the issuer had planned to retire the debt or at times other than when economic savings would result.

Finally, initiating a revenue bond program is more complex and generally requires more time than GO borrowing. Revenue bonds usually have more complex legal, engineering, and trustee calculations; therefore, the related fees are also higher. Projects to be financed typically need a demonstration of engineering and financial feasibility. A revenue bond issue's added cost has important implications for an issuer. In some instances, the benefits of revenue bond financing may not be worth the added cost. A variety of factors, such as total debt, issue size, rating, and bond proceeds, should be considered in making the choice between revenue and GO bonds.

Bond Structure Variations

An issuer can increase the marketability of its bonds by using a variety of bond structures to address specific market conditions. For example, during periods when interest rates are expected to increase, investors are more interested in floating-rate debt; however, this is the time when issuers are interested in fixing rates at current levels. Investment bankers continually try to satisfy both interests by developing new bond structures and instruments that offer acceptable terms to both parties. Historically, such structuring included the development of variable-rate demand obligations (put bonds), floating-rate obligations, original-issue discount (OID) bonds, zero-coupon bonds, and capital appreciation bonds. The issuer needs to be aware that if it selects such variations in order to sell the bonds, it may be accepting more interest rate risk. The issuer should discuss alternative structures in depth with its financial advisor and ensure that it understands both the benefits and the risks it can expect. Characteristics, advantages and disadvantages of several of the newer structures are discussed in the following paragraphs and are summarized in Table 3-1.

Variable-rate demand obligations (tender-option or put bonds). The most significant advantage of variable-rate demand obligations is the interest savings, which can be substantial. Investors are willing to accept a lower rate on these instruments during a period of steep yield curves because the put option provides a way to exit the bonds at par before maturity if an investor needs to do so.

The amount of savings will, however, vary with market conditions, the credit rating of the issuer, and particular put-option features. In addition, the letter-of-credit fee, any insurance premiums, and the cost of funds potentially borrowed against the letter of credit can offset potential savings.

Discount bonds. Discount bonds have two primary advantages for the issuer: lower interest costs and a generally lower overall yield to maturity. There are a number of reasons investors are willing to accept a lower yield and lower interest income from discount bonds. In states where the face value of a discount bond can be used to meet pledging requirements, the same dollar amount can buy considerably more face value of discount bonds than of conventional bonds. Similarly, individual investors can purchase a large par amount of discount bonds at a relatively low cost. Discount bonds are also advantageous during a period of active tax swapping, when they are in considerable demand.

Bonds with extreme discounts, such as zero-coupon or capital appreciation bonds, have additional advantages. First, with these instruments, the investor no longer has reinvestment risk, so these bonds are particularly attractive to investors during periods when interest rates are expected to decline and only lower-yielding reinvestment options are anticipated. Second, zero-coupon bonds appeal to a broad market because they are often purchased by individual investors, for whom a $5,000 bond might be prohibitively expensive, in order to provide money for future educational or retirement needs. Because these bonds are attractive to investors, many investors accept a lower yield to maturity than on a conventional bond. Another advantage to the issuer is that a zero-coupon bond incurs fewer administrative expenses, since there are no semiannual interest payments.

A disadvantage of discount bonds for an issuer is that the issuer must sell a greater par amount of bonds to realize the same amount of bond proceeds. This could present problems if the issuer's legal debt authorization is limited. Another consideration for an issuer is that redeeming discount bonds early is expensive. Exercising call provisions at premium or even at par can raise the yield on the bonds considerably.

Bonds with extreme discounts also have disadvantages for issuers. The issuer must carefully structure appropriate sinking-fund provisions to ensure that requisite funds will be available at maturity to meet balloon payments due at that time. Also, some states prohibit excessive discounting of bonds. In other instances, state or local debt limitations or maximum debt authorizations may inhibit the sale of capital appreciation or zero-coupon bonds. Further, it may be difficult to convince the public that issuing bonds with a par value many times greater than bond proceeds is appropriate.

Bonds with municipal bond insurance. The advantage of using bond insurance or other credit enhancements that will result in the highest possible credit rating is that doing so may enable a low-rated issuer to pay a lower interest rate than its own credit rating would normally allow. The use of insurance may also broaden the market for the bonds because some investors demand bond insurance or a minimum rating.

The disadvantages and risks associated with the use of municipal bond insurance include the fact that the up-front insurance payment is deducted from bond proceeds, reducing the amount of funds available for construction projects. In addition, bond insurance may not be available to all issuers. In order to spread their risk, insurers may have limited underwriting capacity in certain geographic areas or with regard to particular types of bond issues. Insurers also frequently impose restrictions on bond issues they underwrite, including requiring reserve funds, high coverage requirements, or rate covenants.

Bonds with letters of credit. The advantages of using a letter of credit to enhance the rating on a bond are the same as those just discussed for bond insurance. As with bond insurance, LOC support can produce considerable interest savings for an issuer. Also, letters of credit may be a requisite component of such financing techniques as variable-rate demand bonds and commercial paper. A disadvantage of using LOCs is that few banks provide commitments beyond 5 to 10 years, so most issues supported by letters of credit have relatively short maturities or are structured to provide for retirement of the debt before expiration of the LOC. Bonds issued for shorter periods tend to have higher annual principal payments than might otherwise be the case, resulting in a greater financial effect on utility financing. Also, the cost of securing a LOC can add measurably to the ongoing expenses of a bond issue. The fee charged is generally payable on an annual basis and represents a fixed percentage of the amount of bond funds secured by the letter of credit. In addition, utility administrative costs may increase because of requirements to file regular financial reports with the bank issuing the letter of credit.

Bond-, tax-, grant-, and revenue-anticipation notes. The major advantage of issuing BANs, RANs, TANs, or GANs instead of long-term bonds is the lower interest rate. At risk, however, is the issuer's ability to enter the long-term bond market at rates less than current interest rate levels when the notes mature. If rates do not decline measurably, future financing costs may increase. Also, if legislation permits, the temptation to roll over the outstanding notes with another short-term issue may overburden the municipality with short-term debt.

Commercial paper (and tax-exempt CP). The primary advantage of CP for an issuer is the extremely low interest rate borne by CP. Because of its short maturities, superior credit quality, and ready liquidity, CP is a prime investment instrument for money market funds. Because of the strong demand for short-term

Table 3-1 Comparisons of funding alternatives

Funding Alternative	Key Components	Advantages	Disadvantages	Conditions Favoring Use of Alternative
1. General obligation (GO) bonds	Pledge of full faith and credit; tax-supported debt	Strongest form of security to investors; lower interest cost to issuer; lower marketing cost	Must seek voter approval; legal debt capacity limitations	Healthy or growing tax base exists; no debt ceiling concerns; city healthier than enterprise
2. Revenue bonds	Pledge of enterprise earnings (revenues); dedicated stream of funding	Generally requires no voter approval; no draw on debt capacity or tax base: user fee structures	Higher interest cost to bond issuer than with GO bonds; more complex and costly transaction than GO bonds	Enterprise healthier than city; city at or approaching debt ceiling for GO debt; relieve pressure on new taxes
3. Double-barrel bonds or combination bonds	Pledge of revenues backed by pledge of full faith and credit	Additional security; enterprise's shortcomings bolstered by pledge of full faith and credit; pledge of revenues may keep borrowing off debt capacity calculation	Must seek voter approval; legal debt capacity limitations	Enterprise new (no track record) or in some distress; poor market interest in enterprise
4. Tender-option bonds or put bonds	Long-term bonds with tender option in advance of maturity; issue has characteristics of a short-term obligation	Takes advantage of lower short-term interest rates	Requires a backup credit facility to ensure sufficient funds for redemption; savings vary with market conditions; more complex and costly	Best used during period of steep interest yield curve
5. Floating rate bonds or variable rate bonds	Interest rates pegged to market indexes, treasury market, or interest rates; rate adjusted at specified times; rates usually have specified floor and ceiling	Lower interest rates since issuer assumes full interest rate risk; popular with investors in a volatile market	Costly to administer; uncertainty of future debt service needs; higher risk to issuer	Attractive investments when investors expect interest rates to rise

Table continues

Table 3-1 Comparisons of funding alternatives (*continued*)

Funding Alternative	Key Components	Advantages	Disadvantages	Conditions Favoring Use of Alternative
6. Zero-coupon bonds or capital appreciation bonds	Long-term bonds sold at deep discount from par; pay no interest before maturity	Lower interest rates since issuer assumes full reinvestment risk; low administrative cost	Sinking fund must be structured to meet balloon payment at maturity; approval needed for maximum debt; authorization may be difficult to obtain	Less prevalent since the 1986 federal tax law changes; check for any legal restrictions on excessive discounting of bonds in each state
7. Original-issue discount (OID) bonds	Long-term bonds sold with low interest rate at discount below par	Lower interest costs to issuers; lower overall yield to maturity	Greater par amount of bonds must be sold to realize same amount of bond proceeds	Attractive during periods when interest rates are thought to be declining
8. Bonds with municipal bond insurance	Purchase of bond insurance gives issuer triple-A bond rating	Low to medium credit ratings become triple-A, thereby lowering interest expense	Insurance payment deducted from bond proceeds; insurers often impose additional restrictions on issuers	Need to obtain investment-grade rating or improve low- to medium-grade credit; insurance premium less than spread between issuer's rating and triple A
9. Bonds with letter of credit (LOC)	Major banks pledge to meet debt service payments promptly in case of issuer default	Lower interest cost to issuer because of upgrade of ratings	Expense of obtaining a letter of credit; higher administrative costs	Need to raise ratings; creditworthiness of bank exceeds that of issuer
10. Bond-, tax-, grant-, and revenue anticipation notes	Short-term borrowing instruments with maturities shorter than 1 year; bank letter of credit usually serves as security	Lower interest costs to issuer	High transaction costs; high administrative costs; interest rates could rise rapidly; bank may not renew letter of credit in unfavorable market conditions	Should be limited to 10–15% of debt portfolio; often issued in periods when interest rates are high, but are expected to drop
11. Commercial paper	Short-term promissory note with maximum maturity of 270 days (average 30–45 days); continuously rolled over	Considerably lower interest costs to issuer; strong demand by funds and individual investors	Requires letter of credit to ensure cash liquidity; costly to administer; bank may fail to renew letter of credit in unfavorable market	Should be limited to 10–15% of total debt portfolio; typically chosen when interest rates are high, but are expected to decline

securities by funds and individual investors, the interest rates on CP are the lowest of all instruments. At the time this manual was written, tax-exempt CP typically traded as much as one full percentage point less than comparable 1-year financing.

There are, however, a number of disadvantages to CP financing. The initial startup and operating costs of such a program are typically greater than those of alternative short-term borrowing methods, although these costs have decreased significantly in recent years. They include the cost of the backup credit facility, bond counsel fees, and transaction costs associated with administering the program. Consequently, $50 million is the generally accepted minimum issue size. Once CP has been issued, the recurring refunding and sale process that gives CP the characteristics of a continuous securities offering, requires daily staff commitment. Issuers also assume all the risks of short-term borrowing when they issue CP. Depending on market conditions, the maturing issues may not be able to be refinanced with new commercial paper. Another consideration for tax-exempt issuers is that if the backup line is drawn on, the new loan will probably carry a taxable interest rate.

SUMMARY

When evaluating alternative methods of financing capital expenditures, a utility will need to look at the legal and financial environments in which it operates, as well as the advantages and disadvantages of the various forms of financing that are available to it. In making its selection, it will want to consider not only its financial capacity but also the public acceptability of the various alternatives. How the various alternatives will affect the utility's future financial flexibility and how they fit within its long-term financial strategy are also important considerations.

Chapter 4

Going to Market: The Process

External funding alternatives, including the use of debt, are discussed in chapters 2 and 3. This chapter focuses on the issuance of bonds. For the vast majority of public utilities, the most common type of debt used is the tax-exempt bond.

Bond issuance is one step in the implementation of a capital improvement plan and an overall strategic plan. Financing goals and objectives should be set within parameters that enhance a utility's operational effectiveness. At a minimum, goals relating to the following issues should be established at the start of the financing process:

- Financing costs. A budget should be set for the up-front financing costs. Parameters for the annual debt service (principal and interest) may be included within an existing rate or financial planning study.
- Flexibility. Covenants made to bondholders should not restrict the utility from financing future capital improvements, setting rates, or operating the system in accordance with the CIP and the overall strategic plan.
- Amortization period. The final maturity of the bonds should not exceed the estimated useful life of the project.
- Integrity. The financing process should be carried out in a manner that protects the public trust in the utility.

Although the utility may have the assistance of various outside professionals throughout the financing process, the utility's governing body remains legally responsible for the actions taken and information provided to investors. Thus, it is important that utility staff control the process and secure the most cost-effective results. Staff members who understand the overall financing process and can effectively manage the team, including outside professionals, will be more effective agents of the governing authority. This chapter discusses the financing process. The roles and responsibilities of each team member are discussed in chapter 5.

OBTAINING AUTHORITY

Before a utility can issue long-term debt, it must obtain authorization from its governing body, a city council or town governing body, a utility commission, or a regional body. In order to obtain approval, the utility should present

- the capital program
- its plan for funding the capital program
- its plan both for payment of the debt and for utility operation

Often a rate study or financial plan is used to show the governing body how current or new rates will support the additional bonds. In some cases, a cost-of-service study will be necessary. Before the governing body approves the bond issuance and rate study, a public hearing may be required. At the hearing, ratepayers and other interested parties can speak for or against the utility's capital program and proposed rates. If debt is to be secured by a GO of the taxing power of the state or local community, some utilities must obtain approval by the adoption of a referendum of issuance on the local or statewide ballot.

In all likelihood, the engagement of professionals, the bond resolution, and the bond purchase agreement (BPA) will need approval by the utility's governing body. The BPA is often adopted as part of an "award resolution" incorporating details of the financing terms.

In most (but not all) states, publicly owned utilities do not have to file with the state's public utility or service commission. Investor-owned utilities must file with and obtain approval of the state agency and may be required to obtain SEC approval before issuance of the long-term debt.

ASSEMBLING THE FINANCE TEAM

After the governing body has approved the capital improvement plan (or refunding) and has instructed staff to assemble a finance team to implement the financing plan, the internal resources of the issuer should be evaluated. The staff should also evaluate the status of any outstanding commitments or contracts with outside professionals. It is not uncommon for issuers to engage a bond counsel, financial advisor, or both for a period of time, perhaps 2 years, rather than for a specific issue. For the purpose of the following discussion, it is assumed that the issuer has no existing relationship with a bond counsel, financial advisor, or underwriter. However, it is also assumed that the project principals—consulting engineer and feasibility or rate consultant—have established working relationships with the utility.

Since the financial advisor serves the issuer in a fiduciary capacity, it may be prudent to select this professional first for guidance in choosing the bond counsel and underwriter. The selection of the latter two professionals should be based on the results of an open, merit-based request for proposals, as well as a ranking system including the following criteria:

- experience of the firm
- experience of the individuals assigned to the account
- physical location in relation to the issuer
- resources available to service the issuer

- knowledge of and ideas relating to the issuer
- structure and level of compensation

The selection process may include a shortlist of three to five firms based on written responses to the request for proposals. After conducting formal interviews of the short-listed firms, the utility should compile a final ranking for approval by the governing body.

The structure of compensation may be more important than the level of compensation in selecting a financial advisor and bond counsel. Fees based solely on the size of the issue may taint the independence of the advice given. Additionally, expertise and level of service can differ greatly among firms applying for these important positions on the finance team. The utility should compare the structure and level of compensation with those of similar transactions by other issuers.

Before selecting the underwriter, the finance team should determine the type of sale. This subject is addressed later in this chapter.

TIMELINE

Assembling the finance team could take from 60 days to 6 months. This team, once it is assembled, should compile a timetable listing each task to be accomplished, the team members who are responsible, and a corresponding due date for each task through bond closing. A timetable, although it should be flexible, will help the individual finance team members focus on their responsibilities and keep the process moving forward. An example of the process and time frame of issuing bonds is illustrated in Figure 4-1.

Before it works with rating agencies, bond insurers, and potential investors, the finance team must have the following documents in substantially final form:

- bond resolution
- engineer's report
- financial feasibility report
- preliminary official statement (POS)

The bond resolution, engineer's report, and financial feasibility report may be drafted on a parallel basis. If a rate study has recently been completed, preparation of these documents may take from 3 to 12 weeks, depending on a variety of factors, including the size of the utility and proposed capital funding requirements. However, if the rate study and financial plan are outdated, three to four times this length of preparation may be necessary. Since these three documents are the basis for the POS, a working draft of the POS is usually the last document available for review. If the transaction is on a fast track, then the documents may be sent to the rating agencies and bond insurers with some missing data, but they should contain no material inaccuracies.

After a package of documents is sent to the rating agencies and insurers, a formal response may take approximately 10 to 21 days. It may be prudent to offer to meet with these entities and make a formal presentation of the package. After the questions of rating and insurance are answered, the POS can be printed and mailed to investors, commencing the marketing period. Allow for a marketing period of 3 to 14 days for the bond sale. Bond closing usually follows the sale by 7 to 14 days. On the date of the bond sale, the terms, interest rate, price, and amount of transaction are agreed to by the issuer and the underwriter. The actual exchange of bonds for dollars takes place at the closing.

Task Identification Number	Task Description	Duration, *days*
1.	Hold first meeting of finance team to discuss timetable and other matters.	1
2.	Distribute first draft of engineer's report, financial feasibility report, and bond resolution.	20
3.	Hold conference call or meeting for document review.	3
4.	Distribute second draft of engineer's report, financial feasibility report, and bond resolution, as well as first draft of preliminary official statement.	6
5.	Conduct second document review.	3
6.	Send packages to the bond insurers and rating agencies.	7
7.	Make site visits, road trip presentations, or both, to bond insurers and rating agencies.	10
8.	Receive ratings and insurance commitments.	7
9.	Distribute final preliminary official statement draft.	3
10.	Print and mail preliminary official statement.	2
11.	Market bonds.*	12
12.	Price bonds.	1
13.	Award bonds.	1
14.	Print and mail final official statement.	2
15.	Conduct closing.	12
	Total	90

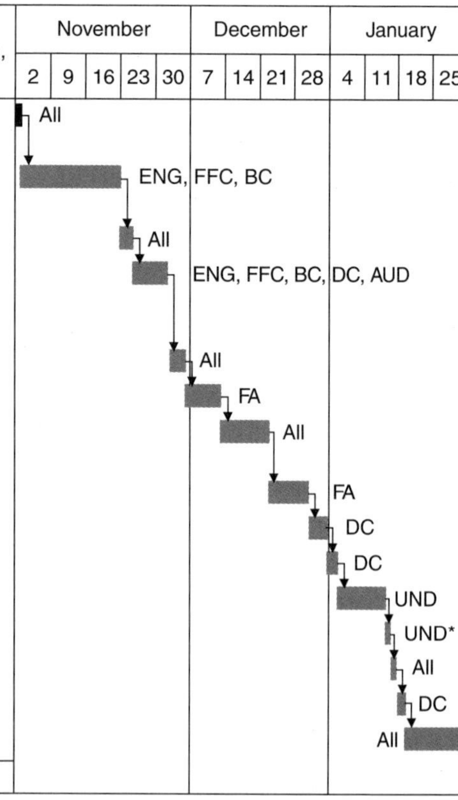

LEGEND
AUD: Auditor
BC: Bond counsel
DC: Disclosure counsel (optional)
ENG: Engineer
FA: Financial advisor
FFC: Financial feasibility consultant
UND: Underwriter *: Underwriter, if sale is negotiated
Task
Progress

Figure 4-1 Sample timeline for bond issuance

Most states have a validation process during which the issuer may file a validation complaint with the state court; this complaint will provide notice and a forum for interested parties to challenge the bond resolution and the issuance of bonds. Basically, this process provides the issuer protection from legal challenges subsequent to the issuance of bonds. Normally the procedure takes 60 to 90 days and can run parallel to other tasks, except closing, once the resolution is available. The issue of validation should be discussed with the bond issuer's own counsel or bond counsel, or both.

PREPARING THE OFFICIAL STATEMENT

The official statement is the document that presents a utility to the underwriter and then to the ultimate investor in the bonds. It provides the following information:

- description of the bonds
- description of the project being financed

- use of the proceeds to be derived from the sale of bonds
- an overview of the operation of the utility, including historical and projected financial statistics and other pertinent legal and tax information
- description of the system, customers, rates, and billings, as well as an organizational chart

The engineer's report, the bond resolution, and the financial feasibility report are the primary sources of information contained in the official statement. The SEC has promulgated certain regulations requiring the issuer to certify that the official statement, as of the date of its publication, does not contain any untrue or misleading remarks or misrepresentations of material facts (in light of the circumstances under which the statement was prepared).

When bonds are sold through a competitive sale, the official statement is put together by the issuer, financial advisor, consulting engineer, financial feasibility consultant, bond counsel, and the utility's independent auditor or accountant. When the sale of bonds is negotiated with an underwriter, the financial team also has a legal responsibility for the quality of information in the official statement and will conduct—usually with the assistance of an underwriter's counsel—a due diligence review of the disclosure information contained in the official statement.

The official statement is prepared in two phases. Most of the work is done in the first stage and results in a POS. This POS, usually printed, is distributed to prospective buyers and to underwriters who, in the case of a competitive sale, are prospective bidders on the bonds. Once the bonds are awarded to an underwriter, the POS is updated with prices received from the sale, including interest rates and yields for each serial and term bond, cost of issuance of the bonds (including cost for the underwriting), debt service information, and any other material information that was not a part of the POS. The final official statement is then printed. The cost of printing and mailing the POS and the final official statement is borne by the issuer.

METHOD OF SALE

Issuers have three methods available for selling bonds:

- private placement
- negotiated sale
- competitive sale

An issuer should review each method in light of financing costs (including interest), level of service, and political considerations.

Private Placement

In a private placement, the issuer sells bonds directly to a small number of sophisticated investors who have the ability to perform their own assessments of the issuer's credit. If the issuer's credit would likely be deemed as noninvestment grade by a rating agency, the issuer might find it beneficial under certain circumstances to consider a private placement.

Since bonds sold through private placement are illiquid to the investor (i.e., the investor is restricted from selling them), the transaction inherently results in a higher effective interest cost. In the mid 1990s, less than 2 percent of the total dollar amount of tax-exempt bonds had been privately placed. Therefore, this chapter will emphasize the competitive and negotiated methods of sale.

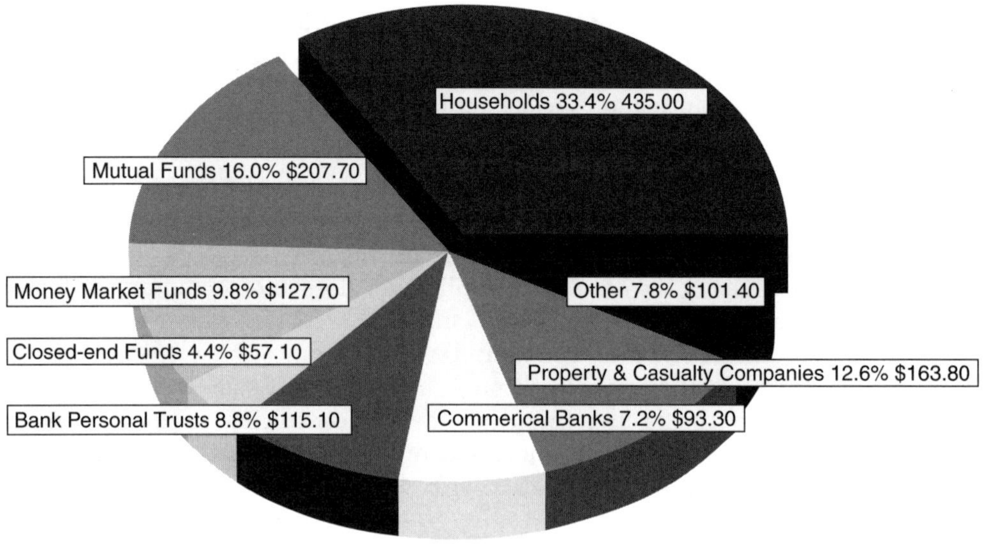

Source: Federal Reserve System, 1995

NOTE: All amounts in billion $US

Figure 4-2 Investor groups' holdings of municipal securities

Bonds sold on a competitive or negotiated basis are publicly offered to a world of potential investors. To the issuer, a bond issue appears to be one large debt obligation with a specific interest rate. However, publicly offered bonds are traditionally sold in $5,000 denominations with different interest rates for each maturity in order to appeal to various investor groups. Therefore, each bond issue is typically purchased by numerous investors. Approximately $1.3 trillion worth of tax-exempt bonds was outstanding in the United States as of 1995, with the percentages of ownership for different investor groups summarized in Figure 4-2.

Negotiated Sale

A negotiated sale involves selecting the underwriting firm or firms in advance of the sale. In a negotiated sale, the underwriters and other professionals assist the issuer in structuring the debt obligation. The issuer negotiates the price for the issue with the underwriters at the time of sale. Before the decision is made to negotiate the sale of bonds, a set of facts should be developed to justify choosing the negotiated form of sale over the competitive sale. For a complex credit, a negotiated issue can be justified based on reasons that include

- a volatile market
- an unusual bond structure
- a refunding of bonds
- an issuer that is in the market infrequently

A financial advisor who has a feel for the market can be helpful to the issuer in identifying facts that could provide justification for a negotiated sale. With a negotiated sale, the underwriter often assists in preparing the official statement and other bond documents used to market the bonds. A disadvantage of a negotiated sale is that the process of establishing the yield on bonds is not as straightforward as in the competitive form of sale. An issuer must evaluate the price offered by comparing the issue against other equivalent issues offered just before the sale of bonds. A financial advisor is often needed to help the issuer evaluate the terms and conditions the underwriters offer. Table 4-1 provides a comparison of conditions favoring either negotiated or competitive sales.

Table 4-1 Summary of conditions favoring either the competitive or negotiated method of sale

Debt Structure Component	Conditions Favoring a Competitive Sale	Conditions Favoring a Negotiated Sale
Security Characteristics		
Pledged revenues	General obligation or strong system revenue	Project-supported revenues
Security structure (for revenue bonds)	Conventional resolution and cash flow; rate covenant and coverage	Unusual or weak covenants; subordinated debt
Debt instrument	Traditional serial and term, full-coupon bonds	Use of innovative structuring; derivative products; structure to attract particular investors (i.e., discount bonds)
Credit Quality		
Rating level	A or better	Below single A
Rating outlook	Stable, A or better	Weak but improving, or under stress
Issuer Characteristics		
Type of organization	Broad-based, general-purpose borrower	Special-purpose, independent authority
Frequency of issuance	Regular borrower in public market	New or infrequent issuer
Market awareness	Active, informed issuer staff	Unsophisticated issuer staff
Investor comfort	Well-known, stable issuer	Issuer experiencing significant financial, legal, or other problems
Market Conditions		
Current trend	Stable, predictable, strong market	Volatile or declining market
Interest rates	Good investor demand; good liquidity	Oversold market; heavy supply
Supply and demand	Light forward calendar; good investor demand	Oversold market; heavy supply
Policy Consideration		
Participation in sale of bonds	Desire to have broad market participation for sale of bonds	Desire to direct business to disadvantaged business enterprises or to local or regional firms
Stimulation of investor interest	Desire to have broad market participation for purchase of bonds	Desire to direct business to local or regional investors

Competitive Sale

The competitive form of sale is the easiest and most straightforward method to sell a municipal debt obligation. It differs from the negotiated sale in that the issuer does not select the underwriting team but rather publishes (1) a notice of sale that specifies, among other things, how bids are evaluated (usually on the basis of net interest cost or a true interest cost) and (2) a bid form that is mailed with the POS to all prospective underwriters. The underwriters then form underwriting teams and submit bids on the bid form at the time and place specified in the notice of sale. The issuer opens the bids at the prescribed time and, after compiling the results, awards the sale to the underwriting team that submits the lowest interest cost. A competitive sale is often used for the following reasons or circumstances:

- The market is stable.
- The bond structure is a traditional one.
- The issuer is recognized in the marketplace.
- The issuer is legally required to hold a competitive sale.

Using this method, an issuer is assured a competitive or low rate on the bonds. In a competitive sale, the bidder has the option of obtaining credit enhancement by purchasing bond insurance. The cost of this insurance can be borne by the underwriter or the issuer as specified in the notice of sale. The underwriting team will use bond insurance if it believes that a lower bid can be produced as a result. Also, some investors are interested in purchasing only bonds of a high credit rating, and this insurance provides such a rating.

UNDERWRITER CONSIDERATIONS

Underwriters enter the bond process in two ways. In a competitive sale, they bid on the bonds at the time of sale. The bid that produces the lowest true interest cost determines who the underwriter will be for the issue. In a negotiated transaction, the issuer actively selects the underwriter for the issue.

Depending on the size of the issue—small, medium, or large—the issuer decides along with the senior underwriter if the issue needs co-managers. A senior manager negotiates directly with the issuer the terms and conditions for the bonds and assists in establishing the best time to sell the issue. The senior manager keeps the books and allocates the bonds to the various managers. The issuer, along with the financial advisor, oversees this activity. The use of co-managers helps in gaining access to various segments of the market that might not be thoroughly covered by the senior manager. Some underwriting firms have the sales strength to access institutional markets, whereas others have access to retail markets. The markets that the issuer is trying to reach determine the choice of co-managers. In some cases—for example, for large offerings or during difficult market conditions—a selling group will be established to provide an additional level of access to the market. Selling group members usually assume none of the underwriting risk and do not cost the issuer any additional money. The selling group members are paid from the management fee or underwriting risk component of the spread, if applicable, that is paid to the managers.

SALE AND CLOSING

The sale or pricing is the most important event in the financing process. This is when interest rates and other terms of the transaction are determined. The underwriter will offer rates and terms that are supported by offers from potential investors. Since the investors hold offers open for only a short period of time, the issuer's staff, with the financial advisor's assistance, should be prepared to give a "verbal award" within minutes after bids are taken in the case of a competitive sale or upon conclusion of discussions in a negotiated transaction. Formal approval of the governing body should take place within 24 to 48 hours of the verbal award. There is a period of time during which all the documents of sale—including legal and tax documents, plus those required by the governing resolution—must be completed. On the prearranged date, these documents are executed and the sale is finalized. In order to avoid last-minute problems, most documents are executed in a preclosing held before the actual date of closing. On the date of closing, the underwriter usually wires the net bond proceeds to the appropriate bank or trust company designated by the issuer, and the issuer simultaneously delivers the bonds to the underwriter.

POSTSALE CONSIDERATIONS

After the sale is completed, the finance team should evaluate the success of the sale in relation to the goals and objectives set in the initial stages of financing. For example, the team should answer the following questions:

- Do the true interest cost and amortization of the debt fit well within the parameters established in the various reports used to educate the governing body and the public?
- Did the issuer retain the flexibility to easily finance future capital improvements?
- Did financing costs stay within estimates and bids?
- Were credit ratings of the issuer retained, downgraded, or enhanced?
- Did the process enhance relations between the issuer and investors?
- Did each team member perform as promised?
- Did the team work well together?
- How can the process be improved?

By documenting the answers to these and other questions, an issuer can enhance the success of future financings. Many issuers also find it helpful to have the finance team continue to meet on a periodic basis in order to improve communication and to better manage the capital improvement funding process. This communication also improves the issuer's ability to monitor the outstanding debt for possible refunding opportunities.

The public offering carried out by the issuer is not the last time the bonds are sold. Initial investors frequently sell bonds to other investors in the secondary market. In order to protect all investors, the SEC adopted continuing disclosure requirements which were made effective July 1995. These requirements call for issuers to update material information contained in the official statement on an annual basis and to provide notice of *any* of the following 11 material events. The financial advisor should be able to tell whether these apply:

- principal and interest payment delinquencies
- defaults for reasons other than nonpayment of interest or principal
- unscheduled draws on debt service reserves reflecting financial difficulties
- unscheduled draws on credit enhancements reflecting financial difficulties
- substitution of credit or liquidity providers, or the failure of these providers to perform
- adverse tax opinions or events affecting the tax-exempt status of the security
- modifications to the rights of security holders
- optional or unscheduled bond calls
- defeasance
- release, substitution, or sale of property securing repayment of the securities
- rating changes

The required information has to be filed with a Nationally Recognized Municipal Securities Information Repository (NRMSIR) in order to ensure its availability to all potential investors. The full finance team should discuss identification of NRMSIRs and other issues of compliance with SEC continuing disclosure requirements early in the process, with the objective of providing adequate disclosure while not unduly burdening the issuer subsequent to the bond closing.

After the transaction has been completed, bond counsel usually provides each finance team member a transcript containing copies of the bond documents associated with the transaction. This serves as an easy reference tool throughout the life of the bond issue.

SUMMARY

Bond issuance is one step in a capital improvement plan. It involves obtaining the authority to issue a bond, assembling a finance team, and compiling a timeline to schedule the tasks. The official statement is prepared by various members of the finance team, depending on the method of sale—private placement, negotiated sale, or competitive sale. During the sale, interest rates and other terms of the transaction are determined, making the sale the most important event of the financing process. After the sale is completed and the documents executed, the finance team should evaluate the sale in relation to goals and objectives established at the beginning of the process.

Chapter 5

Going to Market: The Players

Because so many participants can be involved in the course of a financing, this chapter is dedicated to providing a broad explanation of the roles these individuals can play. It is beyond the scope of this manual to detail specific responsibilities of each member of the finance team. The descriptions that follow identify the participants in the process and offer a synopsis of the responsibilities of each participant as a member of the finance team, the organizational role for each participant, and how the team member may be compensated.

ISSUER PRINCIPALS

The financing staff of the issuing entity should have the responsibility for the bond-financing program. No matter how many outside consultants are retained, the issuer is responsible for all aspects of the program. The lead finance staff is responsible for all financial information, rates and revenues data, and all statistical data (e.g., concerning population growth and economic issues). Because most of the official statement is technical in nature, the inclusion of issuer personnel responsible for operating and engineering areas is critical. After the expertise needed on the financing team is determined, outside consultants should be selected. Potential team members are listed in Table 5-1.

Typically, successful financing is the result of cooperative efforts of the issuer's professional staff working in tandem with consultants representing the legal, financial, and engineering disciplines. This group makes up the finance team—the individuals charged with the responsibility to develop and implement bond financing programs.

The finance team may take different forms depending on the nature of the financing. A bond financing, for example, is the most complicated financing process, particularly when compared to a bank loan. It requires the collective expertise of professionals from many different areas. In the form of a public offering, it necessitates extensive legal documentation, preparation of disclosure materials,

Table 5-1 Potential finance team members

Team Member	Description
Accountant/certified public accountant	Provides most recent audited financial statements; may provide escrow verification
Issuer	Determines need for bonding; selects financing team
Bond counsel	Provides expertise in securities law; represents interests of future purchasers
Issuer's counsel	Reviews local legal matters for inclusion in official statement
Disclosure counsel	Conducts due diligence; may draft the official statement; provides guidance for ongoing disclosure responsibilities
Underwriter's counsel	Reviews all documents; may prepare official statement and other documents of concern to the underwriter
Financial advisor	Provides financial analysis and assistance with the issue; represents issuer
Rating agency	Rates creditworthiness of entity for investors
Credit enhancer	Insures the bonds or provides a letter of credit to back the bonds
Underwriter	Purchases bond and resells to investors
Commercial bank	May offer tax-exempt loans or credit facilities to traditional municipal issuers
Financial feasibility consultant	Develops a financial feasibility analysis of the utility
Consulting engineer	Reviews the operations and capital plans and provides engineer's report
Trustee	Holds funds and is responsible for protecting investors' rights
Auditor	Provides most recent audited financial statements and may provide escrow verification
Financial printer	Prints official statement, notice of sale, and other items as necessary
Registrar	Maintains records of bond ownership
Paying agent	Disperses principal and interest payments to bondholders

selection of an underwriter, and meetings with the rating agencies and bond insurers. Bank loans, on the other hand, are based on proprietary credit evaluation and can be completed on standardized loan documents. Both types of financing are important and require the organization of a finance team to ensure a successful closing. The composition of the finance team is based on the financing requirements. It is incumbent on the issuer and consultants to mobilize the resources necessary to organize an efficient finance team.

LEGAL PRINCIPALS

Legal counsel will fulfill a number of diverse roles in the course of any financing program. Documentation and legal opinions are required on myriad issues ranging from the legal organization of the borrowing entity to the legality of an investor in another state purchasing bonds. Generally, there are three or four attorneys involved in a bond issue, depending on the complexity of the transaction. For example, if a letter of credit is used to provide additional security to bondholders, bank counsel represents the bank's interests.

Bond Counsel

A bond counsel is an attorney or firm of attorneys with specific expertise in securities law. Bond counsel is retained by the issuer and charged with the responsibility to issue an approving opinion that the debt obligation is valid, legal, and binding. To do this, bond counsel prepares documentation authorizing the issuance of debt and determines any conditions with which the issuer must comply. Bond counsel may be involved with various documents, including a bond referendum, state legislation, and notices to bondholders, among others.

Perhaps the most important role for bond counsel is to provide an opinion regarding the tax status of bonds. The primary concern is whether interest paid on the borrowing is exempt from state taxes, federal taxes, or both. Federal law governs the issuance of tax-exempt bonds and provides specific direction concerning the types of projects eligible for tax-exempt financing, as well as refunding (refinancing) limitations. State laws generally focus on a narrower set of issues, such as the purpose of borrowing. It is incumbent on bond counsel to determine the extent of state and federal tax exemption, if any, for governmental borrowings.

Bond counsel's responsibilities continue after the transaction is completed. Since underlying documentation can be complex, bond counsel is frequently called on to provide interpretations and clarifications. This is an important resource when an issuer is monitoring compliance with covenants or conducting an audit.

Issuer's Counsel

The bond counsel works closely with the issuer's counsel, an attorney who may be an in-house staff member or may be retained on a contractual basis. In the course of a debt financing, the primary focus of the issuer's counsel is on administrative and procedural matters. Typical opinions provided by issuer's counsel relate to the legality of organization, the validity of meetings, the process by which bond issuance authorization is obtained, and proper adoption of required documentation, among others. Additionally, the issuer's counsel works closely with bond counsel to review documents and ensure that they are operable.

Disclosure Counsel

With the advent of additional regulations pertaining to disclosure by issuers of tax-exempt debt, a new role of legal counsel has evolved. Disclosure counsel is retained by the issuer and is primarily responsible for conducting due diligence, drafting the official statement, and providing guidance with respect to ongoing disclosure responsibilities. Often the task of drafting the official statement is delegated to the underwriter's counsel (see the next subsection). However, some of the larger issuers are moving toward the use of a disclosure counsel in response to continuing disclosure regulations recently promulgated by the SEC. Bond issuers have recognized that the official statement is their disclosure document and want to have their legal counsel oversee its preparation. Notwithstanding this position, underwriter's counsel still may be delegated the responsibility of preparing the official statement.

Underwriter's Counsel

Much has already been said about the duties of underwriter's counsel, and it is important to acknowledge that this type of attorney represents the interests of the underwriter. In this capacity, underwriter's counsel has certain document preparation responsibilities and reviews all documents and opinions prepared by other types of counsel involved in the transaction. Besides the official statement, underwriter's counsel prepares a purchase contract between the issuer and the underwriter, as well as various underwriting documents and securities filings that are of concern only to the underwriter. Underwriter's counsel is retained by the underwriter, although many issuers may want the opportunity to approve the underwriter's selection.

FINANCIAL PRINCIPALS

An issuer may call on several financial experts to provide assistance in the course of a financing program. Many of these experts possess the same level of financial skills but are distinguished by the role in which they serve the issuer. Financing professionals frequently serve as leaders for the finance team since they coordinate much of the work among attorneys, rating agencies, and bond insurers.

Financial Advisor

A financial advisor may be involved in any type of debt-financing transaction, whether it consists of issuing bonds or undertaking a loan. The financial advisor serves the issuer in a fiduciary capacity and should be retained with a formal contract establishing specific responsibilities. This contract is significant because the issuer frequently acts on recommendations from the financial advisor, and there should be no questions about whose interests the financial advisor represents.

Two different types of financial advisors advise governmental entities on financial matters: (1) the traditional investment bank, which provides both underwriting and financial advisory services, and (2) independent financial advisors, which are firms that provide only financial advisory services and do not participate in any underwriting business.

Issuers can use a financial advisor's services on a broad range of projects. Within the context of this manual, however, the focus is on the development and implementation of efficient financing strategies. To provide these services, the financial advisor helps the issuer to evaluate the creditworthiness of alternative revenue scenarios and covenants. Frequently this involves preliminary discussions with rating agencies and bond insurers to assess risk issues. This assessment helps the issuer understand the ramifications of various structural alternatives and supports the decision-making process.

The financial advisor is capable of assisting an issuer with negotiation and evaluation of various fees and costs, as well as interest rate proposals for notes, bonds, or loan products. An experienced financial advisor is knowledgeable about qualified vendors for such specialty areas as financial printing; trustee, paying agent, or registrar services; and escrow verification.

Compensation for the financial advisor can be determined on an hourly basis, through a transaction fee based on the amount of debt, on an annual retainer, or through a combination of these methods. The issuer should select the compensation method that provides greatest value from both financial and political standpoints.

Underwriter

An underwriter has a different business relationship with the issuer. The underwriter is not contractually bound to act in a manner that serves the interest of the issuer. The issuer must look for ways to ensure that its own interests are protected while also maintaining a good working relationship with the underwriter.

An underwriter provides virtually the same analytical capabilities as a financial advisor. Beyond this similarity, the underwriter acts as an intermediary for an issuer to access capital markets. This market access provides the mechanism for the issuer to raise capital through the bond market by marketing debt securities (generally bonds) to institutional and retail investors. The underwriter negotiates the terms of sale with the issuer on the understanding that (1) the underwriter will use its sales professionals to market the bonds and (2) any unsold amounts will be purchased (underwritten) by the underwriter.

Underwriters generally work on a contingency fee basis, gaining compensation only through the successful sale of bonds. They derive remuneration by purchasing bonds from the issuer at a discount and reoffering such obligations to the public at a higher price. There is rarely a need for an underwriter to invoice an issuer since the underwriter's discount should provide sufficient amounts for professional services, sales commissions to market the bonds, reimbursement of out-of-pocket expenses, and underwriting risk, if any.

Commercial Bank

A commercial bank may also be part of a finance team, depending on the nature of a borrowing. Bond issuance can be an expensive undertaking; for smaller borrowings, securing funds through a bank loan may be more cost-effective. When a commercial bank evaluates a loan application, it reviews many of the same factors considered by a rating agency or bond insurer (see the discussion of credit evaluation principals later in this chapter). However, the one intangible factor for a local bank is knowledge of the local community, including familiarity with development plans. Bank loans can be offered on a tax-exempt basis to traditional municipal issuers. Additionally, certain loans to governmental entities that issue less than $10 million per year can be designated as bank qualified. This is important because it allows a bank to write off a portion of the funding costs, making it possible for the bank to extend a lower interest rate.

Loans are not the only debt product commercial banks offer to municipal borrowers. While such banks traditionally underwrite certain types of debt (e.g., general obligation bonds), they are also becoming more active in providing credit facilities. These facilities range from a letter of credit securing a bond to a line of credit providing a source of funds on which to draw for construction expenditures.

PROJECT PRINCIPALS

Some utility financing may be considered to be project financing. This means that the cash flow produced by a given project is sufficient to meet operational and debt service obligations. To ensure that the project can and will continue to be self-supporting, the financial community requires independent professional studies pertaining to the technical and financial aspects of the project. These services are available through the consultants described in this section.

Financial Feasibility Consultant

A financial feasibility consultant develops a multiyear financial plan for a utility. A feasibility study is intended to project a utility's cash flow based both on historical performance and on expected performance after the project improvements are put into service. The financial community has a vested interest in understanding a utility's ability to meet operational and debt obligations from pledged revenues.

A utility may choose among different types of professionals to prepare a financial feasibility report. For the most part, these consultants fit into one of three categories:

- independent consultants (no specified professional affiliation)
- accounting firms
- engineering firms

An issuer can select any of these three as long as the consultants possess the necessary credentials and experience. For the most part, there are no formal professional standards for the preparation of financial feasibility documents by general consulting and professional engineering firms. However, certified public accounting firms must follow specific guidelines promulgated by the American Institute of Certified Public Accountants. The underwriter and any applicable state oversight body may be able to assist the utility in determining the appropriate party to provide this service. Given the nature of the service provided, the financial feasibility consultant generally reports through the utility's finance director.

A financial feasibility study is actually a compilation of information. The financial feasibility consultant reviews information to confirm that it is reasonable, but requires most, if not all, of the data to be provided by the utility or others on the finance team. Generally, the consultant also assesses the reasonableness of the utility's projections of system growth, water sales, cost trends, and so on. This assessment pertains to operations and construction cost estimates, the service demand forecast, interest rate assumptions for bonds, and other costs. Depending on how the financial feasibility consultant is retained, he or she may or may not be involved in developing projected rates or charges for the utility.

The financial feasibility report usually contains (1) a detailed summary of existing rates, (2) a description of any proposed rate changes and their impact in terms of dollars and percentages, and (3) the consideration of any alternative rate structures in light of customer concerns or utility objectives (e.g., affordability, water conservation). A utility's historical rates, current rates, and trend lines for historical rate increases are often compared with those of other utilities on a local, statewide, or national basis. The impact on and ability to use projected impact fees, a one-time revenue source, to satisfy debt service coverage requirements is often evaluated in this report.

The primary role of the financial feasibility consultant is to prepare projections that illustrate the impact the proposed capital projects and associated debt service will have on the financial position of the utility. This is generally measured by calculating debt service coverage ratios and projecting net cash or fund balance for a multiyear period, typically 5 years.

The financial feasibility consultant may also provide the utility with other, ongoing services required by the bond resolution, including annual determinations of net revenue. Most often through the issuance of a separate report or certificate, often referred to as an "additional bonds" or "parity" certificate, the financial feasibility consultant is often responsible for confirming compliance with bond covenants that—on a pro forma, or forward-looking, basis—evaluate the issuing entity's ability to make future debt service payments. This opinion is based on management's future plans and representations as to estimates and assumptions.

Consulting Engineer

Most financial participants to a bond issue, including the underwriter, credit analyst, and investor, conduct extensive due diligence on each project. This normally begins with an objective assessment of the particular utility by a qualified engineering firm to confirm that the utility is effectively operating and maintaining the system. Such a report is prepared by a consulting engineer with recognized expertise in the water industry. An engineer's report forms an integral part of the disclosure document and is generally reproduced in its entirety as an appendix to the official statement.

The consulting engineer is responsible for (1) providing assurances to the financial community that the utility system is in good condition, that the project improvements are necessary and cost estimates reasonable, that the construction timetable is appropriate, and that permits have been or can be obtained, and (2) providing projections of future operating expenses. An engineer's report ultimately evolves as a compilation of the many detailed reports the consulting engineer submits to the issuer concerning the development of the project. Local officials and utility managers have often used information reported at relatively early stages by a consulting engineer to make decisions on whether to proceed with capital improvement projects. The formal engineer's report describes the plan selected for implementation and how such improvements will benefit the utility.

The engineer's report analyzes and draws conclusions relating to the utility's current and future operations. These results typically include an evaluation of the following aspects to the utility system:

- the age, condition, capacity, and adequacy of water supply sources and water treatment and distribution facilities
- an analysis of historical and projected water demand
- an evaluation of water supply contracts, i.e., what percentage of the total supply a given contract represents for the system and whether the contract terms extend the life beyond the bond issue
- future capital improvements and construction schedules and whether the construction plan is feasible and affordable
- how current and future water supply, treatment, and distribution facilities may impact the environment

During the bond-issuing process, the consulting engineer reports to the utility's management and board of directors. The utility is responsible for soliciting and retaining the consulting engineer's services. Additionally, the consulting engineer may be required through the bond resolution to issue periodic reports or certificates regarding the condition and operation of the utility.

CREDIT EVALUATION PRINCIPALS

As Lamb and Rappaport (1987) state, the concept of rating bonds can be distilled to one question: What is the probability of the timely repayment of principal and interest? The complexity of today's bond market renders traditional methods of credit assessment used by investors nearly impossible. By simplifying the credit rating and ranking of bonds, independent rating agencies have become an essential part of the process investors use in making bond purchase decisions. Rating agencies have developed easily recognized grading systems that measure, on a single scale, bonds from municipal issuers ranging from large state agencies to smaller water districts. If a bond issue has high ratings from an independent rating agency, the investor's willingness to buy increases and the market for the bond broadens, reducing the cost associated with the bond issue. The resulting exposure to a wider segment of the bond market could reduce the bond interest the issuer pays. Standard & Poor's Corporation (S&P's), Moody's Investors Services, Inc. (Moody's), Fitch Investors Service, Inc., and Duff & Phelps are four well-known independent rating agencies with widely used systems of ratings for municipal bonds.

The rating of a given bond is established as the bond is issued and then periodically reviewed until the bond is paid off. This rating expresses the quality of the bond and its credit risk. Key factors used to determine a bond rating are discussed in appendix A.

Although all independent rating agencies use many similar factors, a particular agency may put more weight on one factor in its analysis than another. The issuer should be aware of these differences when selecting an agency.

Trustee

A trustee has several roles in the bonding process and subsequent activities. As a signatory to the bond resolution, the trustee, along with the issuer, has ongoing responsibilities with respect to the bonds. The primary role of the trustee is to act as a fiduciary on behalf of bondholders, that is, to protect the financial interests of the bondholders. The trustee does not get involved with developing the bond structure but does substantially review final documentation to confirm that the document's provisions are administratively feasible. In addition, the trustee's legal counsel reviews and comments on the bond resolution and documents to make sure they meet requirements of the bond resolution, along with applicable legal and tax requirements. At the day of closing, the trustee releases the bonds issued after verifying that the bond proceeds have been wired to specified accounts. After the closing date, the trustee collects, holds, and disburses bond proceeds in accordance with the bond resolution. The trustee also collects, holds, and disburses debt funds. The trustee's responsibilities extend to ensuring that the issuer meets all requirements of the bond resolution during the life of the bonds. A trustee is generally required for revenue bonds or certificates of participation, but may not be required for a general obligation bond issue.

Accountant

For most bonds issued, an independent auditor or accountant provides a copy of the most recent audited financial statements, which are incorporated as an appendix to the official statement. Summaries of this audited financial information are generally included in the body of the official statement. The accountant may review the official statement to ensure that all historical information is accurately presented, is referenced to the complete audited financial statements, and is evaluated in light of supporting schedules and footnotes.

The issuing entity's accountant or external auditors often (but not always) play a role in monitoring ongoing compliance with bond covenants as it relates to satisfaction of debt service coverage requirements. Depending on the exact wording of the bond ordinance or covenants, a certified public accounting firm other than the utility external auditor may fulfill this responsibility.

When an escrow is established, as is the case when bonds are refunded, an accountant is engaged to perform the escrow verification. While the issuing entity's external auditing firm may provide this service, a separate firm is often engaged. The accountant may report to the finance director, or the underwriter may select the verification agent.

SUMMARY

This chapter has discussed the roles participants generally play in capital financing, and Table 5-1 provides a summary of potential team members, including general descriptions of their contributions. Regardless of the participants in the financing and the form that the finance team may take, it is important to remember that the issuer remains responsible for all aspects of the program.

Chapter 6

Special Considerations for Investor-Owned Utilities

Capital-financing considerations for investor-owned utilities differ from those of publicly owned utilities. When choosing a capital-financing plan, investor-owned utilities must consider the availability of equity—both common and preferred—in a generally more rigid regulatory environment, as well as the impact of income and property taxes. Investor-owned utilities, unlike publicly owned utilities, have the opportunity of attracting equity capital in a competitive marketplace. In addition, regulation of investor-owned utilities is normally standardized for similar service utilities within a jurisdiction, which may place restraints on the extent to which a company can use each alternative source of capital.

Current tax laws that make certain financing alternatives attractive to publicly owned utilities do not always offer a similar advantage for investor-owned utilities. However, investor-owned utilities may use tax-exempt financing for water facilities if the state in which they operate is willing to designate some of its capital allocation to the utility.

BUDGETING, PLANNING, AND FORECASTING

Comprehensive forecasts of revenues, expenses, and capital costs assist management with decisions regarding the timing of rate increases and effects on customer bills. An effective planning process is fundamental to positive public relations with customers and regulatory agencies.

As discussed in chapter 1, good financing plans begin with (1) a multiyear capital expenditure budget detailing all major construction projects as well as routine replacements of services, meters, hydrants, and purchase of equipment, and (2) an operating budget for the same time frame.

These combined comprehensive long-range budgets provide the basis to determine

- cash requirements
- targeted capital structure ratios
- appropriate and timely rate increases
- capital-financing options

GENERATION OF FUNDS

Capital expenditures funds are derived from two primary sources: internal and external. Internal funds or cash flows are defined as net income, plus deferred taxes, depreciation, and amortization, less dividends. External funds, as the name suggests, are funds raised from external sources, such as the debt and equity markets and contributions in aid of construction.

Internal Sources of Capital

Utility revenues and the resulting income are obvious internal sources of capital. Other sources of funds available within an investor-owned utility are described in the following paragraph.

Depreciation accruals and retained earnings. Actual sources of funds for a specific project depend on a variety of financial circumstances and timing. Funds for routine replacements and renewals are normally generated internally from depreciation accruals and retained earnings when possible. Because these expenditures do not usually constitute substantial investments, they do not warrant external funding. They typically include replacement of service lines, meters, and hydrants and purchase of minor equipment items, office furniture, and tools.

Depreciation. Depreciation reflects the amount by which the value of property is not restored by current maintenance. This loss in value is primarily due to wear and tear, decay, inadequacy, obsolescence, changes in demand for service, and requirements of public authorities. For one or more of these reasons, each element of utility property must ultimately be retired and, if necessary, replaced.

Investor-owned utilities are permitted to include in revenue requirements an allowance, or provision, for annual depreciation. Depreciation allowances are established at levels that permit systematic amortization of the original cost of investment in various components of utility properties over these components' useful lives. Cash generated by recognition of depreciation charges is available for use to retire debt or to finance capital expenditures. While such expenditures preserve the overall original investment in properties, annual depreciation does not provide adequate capital to replace the investment in times of inflation. If annual depreciation allowances are not retained and reinvested, the original investment may be gradually dissipated. In this case, external financing requirements will be greater than otherwise.

Deferred taxes. Taxes that are charged against current operating results or reflected in the income statement and that are not currently payable to the government are known as deferred taxes. Cash generated by deferring taxes is also a source of funding for capital improvements.

Construction work in progress. Another source of internal funding in certain jurisdictions is construction work in progress (CWIP). An item of major concern to investor-owned utilities today is the methodology regulatory agencies use to recognize costs associated with CWIP. Some regulatory agencies exclude CWIP from the rate base on which utilities are permitted to earn a return until that property is placed in service. In such instances, interest during construction is considered a part of the cost of the property and is capitalized, at which time it becomes a part of the rate base. In such instances it is important for the utility to delay debt repayments until the CWIP is completed and accounted for in the rate base.

Because of the unpredictable and erratic nature of inflation rates and interest costs, investor-owned utilities have sought relief from the administrative burden of charging interest during construction by requesting the inclusion of CWIP in their rate base. This permits earnings at the overall allowed rate of return. Many regulatory agencies now include at least some CWIP in the rate base and permit the utility to earn a return on this work in progress.

Replacement policy. Management should pursue a replacement policy that minimizes total costs and is consistent with rendering safe, reliable water service. Detailed cost and performance records may not always be available for evaluation. Consequently, awareness of the total cost-control objective on the part of operating personnel and the financial manager is beneficial for planning purposes. The annual cost for normal replacements should be incorporated into budgets and cash-flow projections. Because of the long life of a water system plant and equipment, facilities are replaced infrequently; however, delayed replacement can become an additional problem when costs for maintaining a worn-out plant become excessive.

Provisions for replacements of major facilities must also be incorporated into the financial planning process. Projects that constitute a significant commitment of financial resources generally require funding from external sources.

Return on investment. Total return on investment for an investor-owned utility must be sufficient to cover the cost of using various capital resources. Total costs of capital include the interest and issuance cost of indebtedness; dividends and issuance costs of preferred stock; and the cost, including issuance cost, of common equity. The first two components are established contractually and are easily determined from utility records. The cost of equity capital is frequently a controversial subject for the utility, consumer, and regulatory agency. Ideally, the cost of equity capital is equal to the return that investors will require on their market-valued commitments to continue to commit their capital resources to an enterprise. Since investors have different views and expectations about the return to be realized on their common stock investments, determining the market rate of return to attract capital is difficult and subject to special studies that recognize market conditions.

External Sources of Capital

Typically, major capital projects cannot be funded from funds generated in normal annual operations because of the magnitude of expenditures associated with these projects. For major construction projects, external sources of capital (debt and equity issues) must be secured.

Debt. Bonded indebtedness is normally secured by a mortgage of the utility's property. It has the highest lien, or call, on the utility's assets. That is, interest payments on bonded indebtedness must customarily be paid before dividends can be paid on preferred and common stock. In case of liquidation, bonded indebtedness principal is repaid before refunding of stock. From a bond investor's viewpoint, bonded indebtedness is the most secure and least risky form of capital investment in an investor-owned utility. Interest on debt is also deducted for income tax purposes. Accordingly, the cost of this debt is lower than that for preferred or common stock.

Long-term debt. Long-term debt can be taxable or tax-exempt. In most cases, taxable debt is privately placed with an insurance company or other large institutional investor. From time to time, utilities may prefer to issue debt by means of a public offering to maintain relations with retail investors. Another option for raising taxable debt is to set up a medium-term note (MTN) program. MTNs allow companies to vary the size and maturity of an issue just days before pricing; in that way they are very flexible. Because of the substantial setup costs, MTNs are suitable only for larger issuers (i.e., issues of $75 million or more).

Investor-owned water utilities in most states of the United States also have access to tax-exempt financing. On average, the cost of a long-term (i.e., 30-year) tax-exempt issue is approximately 2.0 percent lower than that of a comparable taxable issue. Utility commissions and rate-case intervenors (large users and public advocates) encourage this method of financing because of the substantial cost savings over taxable debt. Refer to the "Legal Factors" section in chapter 3 (page 23) for more detail.

Tax-exempt bonds are generally secured by a first mortgage, although some utilities that do not have a large amount of secured debt have successfully issued debentures. Since 1990, most long-term fixed-rate issues have been marketed with bond insurance, which can further decrease the cost of financing.

Short-term debt. Typically in the form of variable-rate bank lines of credit, short-term debt is used to finance capital projects during construction. Water utilities usually replace these short-term lines with long-term fixed-rate debt or equity, or both, as soon as the completed projects are ready to be placed in service.

Equity. Equity may be provided by issuing common or preferred stock.

Common stock. Two primary objectives of water companies needing to issue common stock to raise capital are (1) to avoid diluting earnings per share and (2) to have funds available to pay common dividends. As a result, to offset the dilutive effect of issuing new shares, most common stock issues occur in conjunction with an acquisition or a rate case that increases earnings. Many companies also have dividend reinvestment plans or customer stock purchase plans or both, that can satisfy moderate capital requirements throughout the year.

Preferred stock. Preferred stock is the second-least-risky type of investment. Its dividends and, in case of liquidation, its principal are payable before such payments are made to holders of common stock. This hybrid of debt and equity is generally considered comparable on a risk basis to a junior lien. Thus, its cost is higher to the utility than debt but not as high as common stock.

Contributions in aid of construction. In certain cases, external funding for construction is obtained by direct contributions from those customers who benefit from the construction. For example, developers usually pay for main extensions to housing developments; these payments may be accounted for in the form of contributions in aid of construction or as customers' advances for construction.

Contributions in aid of construction represent any cash, services, or property permanently contributed by customers for construction purposes. Customers' advances for construction also represent contributions from customers; however, the utility and customer enter into an agreement that the utility will refund part or all of the construction cost to the customer over a specified time period. Any unrefunded balance remaining at the end of the agreement period is transferred to the contribution-in-aid-of-construction account.

Although it is possible for an investor-owned utility to benefit in the short term from customer growth (via operating revenues) while not directly financing any significant additional capital investment, this is normally not the case, particularly in the long term. Also, for rate-making purposes, a contributed plant is normally excluded from the rate base. In such cases, the utility cannot recover an annual return on those dollars. Furthermore, regulators do not usually allow recovery of annual depreciation on the contributed plant. Over time, the utility is confronted with replacing the contributed plant and needs to increase water rates to reflect the need for additional capital requirements in long-term financial planning.

ATTRACTING CAPITAL

The primary objective of investor-owned water companies is to attract capital at the most cost-effective rates in order to minimize the impact on the customers' water service rates. In order to attract capital, regulated utilities must maintain vigilance of their revenue needs and file for rate relief as frequently as necessary through evaluating long-range budgets. A utility that shows itself to be progressive in maintaining adequate rates and return on capital will be more successful in the capital marketplace. In order to attract capital at minimum cost, the utility must also use a balanced capital structure. The most balanced and efficient capital structure combines debt and equity capital.

Appropriate Capital Structure

The age-old question for investor-owned utilities is: What is the optimal capital structure? The debt–equity mix for any company is influenced by the company's individual circumstances, such as the intensity of its capital expenditure program, the regulatory environment, and the impact of the type of financing on the customers' water service rates. The average capitalization in 1996 for the 14 largest investor-owned water utilities is shown in Table 6-1.

Rating Agency Considerations

A utility's size often determines the prudence of obtaining a rating from one or several rating agencies. Seeking a rating from all the rating agencies, or even just one, is time-consuming and can be costly. Frequently, the utility's investment banking firm will provide advice on the economics and the prudence of obtaining a rating. See appendix A for further details on rating agencies.

Regulatory Approvals

Investor-owned utilities must seek regulatory approval of any long-term financing. All of the financing documents, as well as testimony and exhibits justifying the purpose and selection of the particular financing, are required to be filed with the governing regulatory state agency (public service or utility commission) for approval before the actual financing. The financing's prudence, purpose, and impact on the utility's capital structure must be demonstrated and justified to the regulatory agency to gain the needed support. Many regulatory agencies have their own guidelines for what they consider an appropriate capital structure for companies under their jurisdiction. Refer to the section entitled "Obtaining Authority" (page 36 in chapter 4) for more detail.

Table 6-1 Average capitalization for the 14 largest investor-owned utilities in the United States

Type of Capital	Maximum, %	Average, %	Minimum, %
Long-term debt	60	50	33
Preferred stock	4	2	0
Common stock (including retained earnings)	65	48	35

Source: C.A. Turner Utility Reports, 1996.

Regulatory Lag

A major hurdle in keeping a water utility attractive to potential investors is regulatory lag—the lag between the time when additional revenue is needed and the time when the regulatory agent grants rate relief. While state regulatory agencies have done much in recent years to accelerate consideration of rate relief requests and to alleviate the impact of lag, a degree of lag cannot be practically avoided. Therefore, regulated utilities must maintain vigilance over their revenue needs and file for rate relief as frequently as necessary. A utility that shows itself to be progressive in maintaining adequate rates will be more successful in the capital marketplace.

To determine revenue needs, most regulatory agencies require a rate filing to be based on statistics that show historical test year expenditures but do allow for modifications for known and measurable changes. Capital markets, on the other hand, look at the future likelihood of repayment. From the perspective of innovative financing, projected test years can allow a utility to be better prepared to develop a financial plan that will be meaningful to both regulators and capital providers.

OTHER CONSIDERATIONS

Other factors determining the attractiveness of a water utility's stock are the dividend yield, the company's record of dividend increases, the dividend payout ratio (which indicates the future growth potential of the dividend), and potential earnings growth of the utility.

Balancing Investor and Customer Requirements

The interests of stock and bond investors must be balanced with the interests of a utility's customers. Investors need to feel secure that the utility has a sound financial position and that a reasonable return will be earned on their investment commensurate with the risk associated with that investment. Customers, on the other hand, are interested in receiving reliable service at reasonable rates. Customers must be informed that the utility's ability to continue to provide reliable service depends on its maintenance of both a sound financial position and a reliable physical plant, which are necessary to attract needed capital at reasonable rates.

Investor Relations Program

One of the most carefully monitored items among water companies is the price of a company's stock. An active investor relations program is integral in maintaining a fair stock price relative to industry peers. Individual or retail investors own approximately 80 percent of a typical water utility's common stock. Primary reasons for investing include attractive current income from dividends, dividend safety (timing and size), and good prospects of future dividend increases to protect purchasing power over time.

The most successful type of investor relations program provides timely information to the financial community and potential investors so that they are able to make informed investing decisions. Potential stock and bond investors use a number of financial and operating ratios and indexes when analyzing the financial soundness of an investor-owned utility. These tools include projections of future rates of return, earnings coverage of interest costs and preferred stock dividends, and various ratios of operating expenses and revenues to each other and to plant investment.

SUMMARY

It is important to remember that options available for financing differ between publicly and investor-owned utilities. This chapter has discussed considerations specific to investor-owned utilities, including the planning process, internal and external sources of capital, attracting capital, and balancing the interests of both customers and investors.

Appendix A

Rating Agency Requirements

This appendix discusses the purpose of credit-rating agencies, how rating agencies are compensated, their basis of competition, and the rating process. It also examines what a water utility can expect if it decides to request a rating on its debt. Throughout this appendix, the water utility is referred to as the issuer.

WHAT IS A RATING AGENCY?

Rating agencies—such as Duff & Phelps, Fitch Investors Service, Moody's, and Standard & Poor's—are organizations that evaluate and publish opinions with respect to the investment quality of securities. Their ratings have traditionally focused on the likelihood of timely repayment of principal and interest, although some of the organizations have begun to assign a second rating that evaluates the market risk or volatility of certain securities. Rating agencies do not audit financial or other information provided to them, and they do not express any opinion as to the appropriateness of any security for any particular investor. Although these organizations have no official authority, each has earned the trust of the public through many years of operation in the credit evaluation arena. Their opinions are generally viewed as independent and objective assessments of the relative creditworthiness of the security being rated.

A rating agency's opinion is expressed in a report that discusses the issuer and the security being rated. The opinion is also summarized symbolically by one or more letters, such as AAA or B. The rating symbol assigned to the security usually appears on the cover of the final official statement and is widely communicated verbally by sellers of the security and in the financial press. Each agency standardizes its opinions so that comparisons can be made of the relative creditworthiness of securities it has rated. However, while the ratings assigned by different rating agencies are often very similar, they are not identical. Just as people's opinions often differ, so do rating agency opinions. Different ratings may be due to different assessments of risks, the placement of more or less emphasis on various factors, or different opinions about the future impact of probable events. Definitions of long-term debt ratings used by the four organizations mentioned earlier appear in Table A-1. A financial advisor should be aware of and able to explain any differences between the philosophies and analytical processes of the major agencies.

Credit ratings generally are grouped into two broad categories: investment grade and speculative. Many investors such as state and local governments, are permitted to purchase only investment-grade securities.

In addition to the ratings shown, an agency may place a plus (+) or minus (–) sign or a number (e.g., 1 or 2) with the rating symbol to indicate the relative position of a credit within the rating category. Some agencies may also indicate that the rating is conditional. A conditional rating is generally used when the opinion of credit quality is based on a repayment source that depends on the completion of some act or the fulfillment of some conditions—for example, bonds secured by earning or rentals from a project under construction. Agencies may, from time to time, indicate that they are reviewing a rating; they may also indicate whether they anticipate that a rating may be raised or lowered as a result of the review. If an issuer fails to provide sufficient information, ratings may be suspended or withdrawn. Because all ratings are security specific, ratings are withdrawn when an issue matures, is called, or is refinanced.

RATING ECONOMICS

Rating agencies are for-profit organizations. They rate securities, generally at the request of the issuer, for the benefit of the investment community. Some agencies may occasionally issue a rating without having been requested to do so. At the time this manual was drafted, however, that practice was controversial because access to issuer information in such cases is much less extensive.

Full rating reports are available to investors by subscription. Rating agencies compete for subscriptions on the basis of quality of analysis, access to information, and price. Because investors may not have either the time or the in-house capability to perform a thorough credit analysis, they frequently rely on rating agency evaluations. Rating agencies are often organized by specialty, with designated groups of analysts covering such areas as public finance, industrials, financial companies, and sovereign credits. Within this structure, there will probably be another level of specialization. It may be by type of issuer (e.g., airports, utilities, housing) or by geographic region. Because the quality of the analysis varies with the skill and experience of the rating agent assigned, investors often have a preference for a certain organization's ratings in a particular specialty area. Most investors do not subscribe to all the services.

When it requests a rating, an issuer is charged a fee by the agency and may also be assessed an annual fee while the securities are outstanding. The issuer evaluates the economics of whether to obtain a rating based on the anticipated cost of financing with and without a rating. Because many investors buy only rated securities, having a rating broadens the potential pool of buyers. A broader base of buyers should increase demand for the security, thereby lowering the cost of financing. For issues of a very small dollar amount, however, the cost of the rating may be greater than the interest savings, especially if there are local investors who are familiar with the issuer.

In deciding whether to get one or more ratings and in choosing which agencies to engage, the issuer often looks at two factors. One factor is which rating agencies are most frequently used by probable buyers of the security. The other factor is the probability of getting the desired rating from a particular agency given the differences in the credit factors emphasized by the various agencies.

Table A-1 Rating definitions used by four well-known rating agencies

Rating	Description
Duff & Phelps*	
AAA	Highest credit quality. The risk factors are negligible, being only slightly greater than for risk-free US Treasury debt.
AA	High credit quality. Protection factors are strong. Risk is modest but may vary slightly from time to time because of economic conditions.
A	Protection factors are average but adequate. However, risk factors are more variable and greater in periods of economic stress.
BBB	Below average protection factors but still considered sufficient for prudent investment. Considerable variability in risk during economic cycles.
BB	Below investment grade but deemed likely to meet obligations when due. Present or prospective financial protection factors fluctuate according to industry conditions or company fortunes. Overall quality may move up or down frequently within this category.
B	Below investment grade and possessing risk that obligations will fluctuate widely according to economic cycles, industry conditions, or company fortunes. Potential exists for frequent changes in the rating within this category or into a higher or lower rating grade.
CCC	Well below investment grade. Considerable uncertainty exists as to timely payment of principal, interest, or preferred dividends. Protection factors are narrow and risk can be substantial with unfavorable economic or industry conditions, or with unfavorable company developments.
Fitch's*	
AAA	Investment grade; highest credit quality. Exceptionally strong ability to pay interest and repay principal. Unlikely to be affected by reasonably foreseeable events.
AA	Investment grade; very high credit quality. Ability to pay interest and repay principal is very strong, although not quite as strong as with bonds rated *AAA*.
A	Investment grade; high credit quality. Ability to pay interest and repay principal is strong but may be more vulnerable to adverse changes in economic conditions and circumstances than for bonds with higher ratings.
BBB	Investment grade; satisfactory credit quality. Ability to pay interest and repay principal is considered adequate. Adverse changes in economic conditions and circumstances are more likely to have adverse impact on these bonds and, therefore, impair timely payment. Likelihood that the ratings of these bonds will fall beneath investment grade is higher than for bonds with higher ratings.
BB	Considered speculative. Ability to pay interest and repay principal may be affected over time by adverse economic changes, but business and financial alternatives can be identified, which could assist in satisfying the bond's debt service requirements.
B	Highly speculative. Currently meeting debt service requirements, but the probability of continued timely payment of principal and interest reflects a limited margin of safety and the need for reasonable business and economic activity throughout the life of the issue.

NOTES: Underline in this table indicates that the security is considered investment grade.

* Plus (+) or minus (–) may modify the ratings from *AA* to *CCC* to show relative standings within the major ratings categories.

† Moody's applies numerical modifiers 1, 2, and 3 in each generic rating classification from *Aa* to B. The modifier 1 indicates that the issue ranks in the higher end of its generic rating category; the modifier 2 indicates a mid-range ranking; and the modifier 3 indicates that the issue ranks in the lower end of its generic category.

Table continues

Table A-1 Rating definitions used by four well-known rating agencies *(continued)*

Rating	Description
CCC	Identifiable characteristics that may lead to default. Ability to meet obligations requires advantageous business and economic environment.
CC	Minimal protection. Default in payment of interest, principal, or both seems probable over time.
C	Imminent danger of default in payment of interest or repayment of principal.
DDD-D	In default on interest payments, principal repayments, or both. Extremely speculative. Should be valued on the basis of ultimate recovery value in liquidation or reorganization of the obligor.
Moody's*	
Aaa	Best quality. Smallest degree of investment risk. Interest payments protected by a large or exceptionally stable margin; principal is secure. Changes that can be visualized are unlikely to impair the fundamentally strong position of such issues.
Aa	High quality by all standards. Margins of protection may not be as large as in *Aaa* securities, or fluctuations of protective elements may be of greater amplitude. There may be other elements present that make the long-term risks appear somewhat larger than in *Aaa* securities.
A	Has many favorable investment attributes. Considered upper-medium-grade obligation. Factors giving security to principal and interest are adequate, but elements may be present that suggest a susceptibility to impairment some time in the future.
Baa	Medium-grade obligations; i.e., neither highly protected nor poorly secured. Interest payments and principal security appear adequate for the present; certain protective elements may be lacking or may be characteristically unreliable over any great length of time. Has speculative characteristics.
Ba	Has speculative elements. Future cannot be considered as well assured. Protection of interest and principal payments may be very moderate and not well safeguarded during both good and bad times over the future. Uncertainty of position characterizes bonds in this class.
B	Lacks characteristics of a desirable investment. Assurance of interest and principal payments or maintenance of other terms of the contract over any long period of time may be low.
Caa	Poor standing. May be in default or there may be elements of danger with respect to principal or interest.
Ca	Speculative to a high degree. Often in default or showing other marked shortcomings.
C	Lowest-rated class of bonds. Extremely poor prospects of ever attaining any real investment standing.

NOTES: Underline in this table indicates that the security is considered investment grade.

* Plus (+) or minus (–) may modify the ratings from *AA* to *CCC* to show relative standings within the major ratings categories.

† Moody's applies numerical modifiers 1, 2, and 3 in each generic rating classification from *Aa* to B. The modifier 1 indicates that the issue ranks in the higher end of its generic rating category; the modifier 2 indicates a mid-range ranking; and the modifier 3 indicates that the issue ranks in the lower end of its generic category.

Table continues

Table A-1 Rating definitions used by four well-known rating agencies *(continued)*

Rating	Description
Standard & Poor's*	
AAA	Highest credit quality. Capacity to pay interest and repay principal is extremely strong.
AA	High credit quality. Very strong capacity to pay interest and repay principal. Differs from the highest-rated issues only in small degree.
A	Strong capacity to pay interest and repay principal. Is somewhat susceptible to the adverse effect of changes in economic and other circumstances.
BBB	Adequate capacity to pay interest and repay principal. Adverse economic conditions or changing circumstances can lead to insufficient capacity to pay interest and repay principal.
BB	Has less chance to default than some other bond issuers. Ongoing uncertainties or exposure to adverse business, financial, or economic conditions could lead to insufficient capacity to pay interest and repay principal.
B	Can meet current interest and principal payments, but adverse business, financial, or economic conditions will likely impair capacity or willingness to pay interest and repay principal.
CCC	Has a currently identifiable vulnerability to default. In the event of adverse business, financial, or economic conditions, it is not likely to have the capacity to pay interest and repay principal.
CC	This rating applies to debt subordinated to senior debt that is assigned an actual or implied *CCC* debt rating.
C	This rating applies to debt subordinated to senior debt that is assigned an actual or implied *CCC* debt rating.
C1	This rating applies to income bonds on which no interest is being paid.
D	In payment default S&P's uses this rating when interest payments or principal payments are not made on date due even if applicable grace period has not expired, unless S&P's believes such payments will be made during the grace period. This rating applies to issuers filing a bankruptcy petition if debt service payments are jeopardized.

NOTES: Underline in this table indicates that the security is considered investment grade.

* Plus (+) or minus (–) may modify the ratings from *AA* to *CCC* to show relative standings within the major ratings categories.

† Moody's applies numerical modifiers 1, 2, and 3 in each generic rating classification from *Aa* to B. The modifier 1 indicates that the issue ranks in the higher end of its generic rating category; the modifier 2 indicates a mid-range ranking; and the modifier 3 indicates that the issue ranks in the lower end of its generic category.

THE PROCESS

The issuer must apply to the agency for a rating and must furnish information the agency needs to form an opinion about the credit quality of the debt to be sold. The rating request is entered on a calendar and assigned a priority based on the anticipated sale date. The application is then assigned to an analyst. The analyst reviews the material submitted, becomes familiar with the issuer and the security structure, calculates various financial ratios, and may, if necessary or helpful, meet with the issuer to clarify information. The analyst's evaluation and rating recommendation are condensed into a report presented to the analyst's manager for review. The analyst and manager then present their conclusions to a rating committee. The rating committee's role is to assign a rating that reflects the creditworthiness of the security relative to all other securities rated by the agency. This requires familiarity with issues and concerns that cross industries.

The committee's rating opinion is released first to the issuer and then to the public. The release is followed shortly by distribution of the full credit report. The issuer may or may not have an opportunity to review the report before its release.

The assignment of the rating needs to be closely coordinated with the issuer's calendar, especially in the case of a competitive sale. The issuer's financial advisor is frequently a key player in coordinating activities surrounding the rating process. See chapters 4 and 5.

As part of the rating process, the issuer generally agrees to provide continuing information to the rating agency. Based on its evaluation of the information provided and other generally available information, the rating agency may from time to time confirm or change the rating on outstanding securities. Changes are communicated to subscribers and are often reported in the financial press.

WHAT IS CONSIDERED WHEN A RATING IS ASSIGNED?

When evaluating a debt issue, the rating agency is most concerned with the likelihood of timely repayment of principal and payment of interest. It will ask the same questions any lender would consider: Will the borrower want to repay the debt? Will the borrower be able to repay the debt? What will happen if the borrower doesn't repay the debt? These key questions are also referred to as the three "C's" of credit: character, capacity, and collateral. Detailed variations of these questions are listed in Table A-2.

WHAT WILL THE AGENCY REQUIRE?

In order to answer the questions that arise when evaluating the security, the rating agency needs to review various types of information, both financial and nonfinancial. The most commonly requested documents are listed in Table A-3. Some of the documents are provided only the first time an issuer obtains a rating; others are provided or updated annually. Still others are only available a few weeks before the sale. The issuer's financial advisor should help determine which documents are relevant and when they will be needed. Some advisors also provide assistance with gathering information.

Communicating With the Rating Agency

It is essential that a small, first-time, or infrequent issuer discuss the rating process with its financial advisor. The financial advisor should be able to provide insight into which agencies would be the best fit for the issuer and to help assess what rating will probably be assigned. The financial advisor may already know the industry analyst at each of several rating agencies. Although it is common for the financial advisor to make initial contacts with appropriate analysts, a large, frequent issuer may feel comfortable performing these tasks in-house.

Communication with the rating agency takes multiple forms. The first is through documents that the issuer provides to the agency. The second is via an informal question-and-answer process as the analyst reviews the documents. The

Table A-2 Rating agency considerations

Category	Relevant Questions
Character: willingness to pay	• Is this a reputable issuer? Is its management reliable? Does it have a history of meeting its obligations when due? • What is the purpose of the borrowing? Is the project being financed essential or nonessential to the operation of the organization?
Capacity: ability to pay	• Will the cash flows from the project or the net operating revenues of the organization be sufficient to service the debt? Are they legally available to service the debt? Can additional debt be issued? If so, is there any limitation on the amount or any preference for holders of this issue? How will additional debt affect the organization's ability to make timely payments of principal and interest on this security? What intergovernmental relationships are in place? How reliant are the various governmental units on intergovernmental transfers? • Is management competent? Do key personnel have the experience needed to operate the organization in an efficient manner? How long have the key players been in their positions? Is the organization stable or in chaos? Is past performance likely to be representative of future performance? • Is the water supply adequate? What population growth is projected in the service area? What is the condition of the physical plant? Are there any regulatory compliance issues that will require major capital expenditures? How large is the capital improvement plan during the next 5–10 years? How does the organization plan to finance its capital investments? • What rate structure is in place? What is the rate-setting philosophy and process? How easy is it to obtain rate increases? Who has control and who can make decisions? What is the organization's relationship with any rate regulatory authorities? What are the service area demographics? What is happening with employment and income in the service area? Can the population afford rate increases?
Collateral: security	• Have revenues been pledged to this issue? Is there a commitment to raise rates as necessary to service the debt? If revenues are insufficient, are property taxes available for debt service? What constitutes a default? In the event of default, has any property been pledged? If so, what is the expected amount that could be raised by forced sale of the pledged property? In the event of default, who will make decisions? Is there an indenture? Who is the trustee? • What legal protections do security holders have? What commitments have been made with respect to disclosure of future information that may affect the security?

financial advisor handles a great deal of this communication. It is essential for the advisor to know the organization well or at least to have access to the appropriate issuer staff to obtain timely and accurate clarification of points arising during the document review process. The more unusual or complicated a security, the more time the rating agency will need to review materials, and the more likely it is that questions will arise during the review. More time and clarification will also be required to review materials from first-time or infrequent issuers than from an issuer with whom the analyst is familiar.

A third form of communication with the rating agency is the formal meeting. Meetings give the analyst a chance to assess management's competency, philosophy, and character. A meeting also gives management a chance to tell its story and to explain complex issues and strategies. Top-level financial and operating managers of the issuer should be involved in the presentation, as should the financial advisor. The

Table A-3 Documents frequently requested by a rating agency during an evaluation

Information Category	Documentation Requested
General	Description of the security
	Preliminary official statement
	Bond ordinance, or resolution, or indenture, plus all supplemental documents
	Bond counsel opinion
	Supply and demand information (historical and projected)
	System description and capacity
	Service area economics and demographics
	Largest customers by revenue and volume
	Largest employers
	Labor relations history and labor contract summaries
	Summary of community involvement and sentiment
	Environmental or regulatory compliance status and strategies
	Engineering report or feasibility study
Financial	Current year budget
	Capital and financial plans or projections
	Rate requirements or rate study (3 to 5 years historical and projected)
	Comparison of rates to those in surrounding communities
	Financial reports based on generally accepted accounting procedures for 3 to 5 years
	History of meeting budgetary and capital plan goals
	Financial ratios

meeting can be structured as a phone conference, or it can be an in-person visit held either at the rating agency's office or at the issuer's facility. A meeting at the issuer's facility has several advantages.

- More of the staff can be involved.
- A facility tour can be used to demonstrate good operating practices.
- Officials who may be unable to commit the time to travel to out-of-town meetings may be able to participate for a portion of the meeting.

A site visit does require more of the analyst's time, however, so it may not be reasonable to expect every rating to involve an in-person meeting at the issuer's facility.

In preparing for a meeting, the issuer should focus on presenting its best side, but it must also be open and forthright in addressing any potential issues or concerns the analyst may have. Because one of the purposes of the meeting is to give the analyst an opportunity to assess the character and integrity of management, trying to hide problems is not a good idea. The meeting will often be structured as an issuer presentation followed by a period during which the analyst asks questions. Color charts, graphs, booklets, and summaries that the analyst can take away from the meeting are often useful. Be prepared to talk about the following issues:

- why the securities are being issued
- how the proceeds will be used

- the organization's financial status and projections
- current operating statistics
- future strategies
- economic and demographic conditions in the service area

It is also helpful for the issuer and the rating agency representatives to agree on a meeting agenda in advance. The agency representatives will want to be sure that adequate time will be allotted for their questions during or after the issuer's presentation so that no essential information will be missed because of time constraints.

An effective communication strategy will involve continued contact after a rating has been issued. Future ratings applications will be less traumatic if the analyst is familiar with the organization and has respect for its management. Material changes in the organization should be communicated to the analyst in between scheduled rating visits, and the rating agency will also want to receive regular financial reports. Some analysts also like to receive the organization's news releases and other public communications. An issuer should discuss with the financial advisor and the rating agency analyst the preferred level of communication between rating meetings

ADDITIONAL INFORMATION

Each of the rating agencies has materials prepared to help issuers and investors understand the role of the rating agency, the agency's rating systems, and its financial analysis process. Those materials can be obtained directly from the agency, or they may be available from the financial advisor. Other organizations, such as the Government Finance Officers Association and the Treasury Management Association, have published material that may be useful to those who need to know more about the rating process. A listing of current publications can be obtained directly from those organizations. Many of the larger underwriters and financial advisory firms also publish booklets that contain useful information.

SUMMARY

A water utility, like other organizations that issue debt, may want to have a major credit rating agency publish an opinion about the credit quality of its debt. A utility that decides to obtain a rating will need to decide, with the assistance of its financial advisor, which agency is most appropriate for the organization and whether to obtain more than one rating.

The utility should be prepared to furnish information to the rating agency about the security to be sold. In addition, the rating agency will probably want to receive information about the utility's legal structure, its financial status, its operations, and the economic environment in which it operates. After the security is sold, the utility is expected to provide certain updated information as long as the security remains outstanding.

Glossary

accelerated depreciation Use of depreciation methods that amortize the cost of an asset at a faster rate than under the straight-line method. The three principal methods of accelerated depreciation are (1) sum of the year's digits, (2) double declining balance, and (3) units of production.

accountant A financial professional who specializes in accounting. In connection with a bond issue, an accountant often provides the most recent audited financial statements and may provide escrow verification.

accrual basis A basis of accounting in which revenues are recognized when earned and expenses are recognized when they become liabilities for benefits received. Receipt of revenue or payment of the expenditure may take place, in whole or in part, in another accounting period.

additional bonds test A requirement in an indenture that additional bonds not be issued unless historical and projected revenues indicate there is sufficient revenue to avoid dilution of coverage on outstanding bonds.

ad valorem tax A state or local tax based on the assessed value of real or personal property.

advance for construction An advance that may be refundable either wholly or in part and that is made to the utility in order to fund construction. When all potential claims for refunds have been settled, the balance, if any, is treated as a contribution in aid of construction, under generally accepted accounting principles.

advance refunding bonds Bonds issued to replace an outstanding bond issue before the date on which the outstanding bonds become due or callable. Proceeds of the advance refunding bonds are deposited in escrow with a fiduciary, invested in US Treasury bonds or other authorized securities, and used to redeem the underlying bonds at maturity or on the call date and to pay interest on the bonds being refunded or on the advance refunding bonds.

ancillary charge A separate charge for ancillary services that is not included in costs for general water service. In providing water service, a utility must often perform these types of services, which often benefit only the individual customer using the services and have no systemwide benefit. Such services might include: account activation, connect and disconnect activities, and meter testing, to name a few.

authority bonds Bonds payable from the revenues of a specific authority, such as a water or sewer authority. Since authorities usually have no revenues other than charges for services, their bonds are ordinarily revenue bonds.

availability charge A limited-use charge made by a water utility to a property owner between the time when water service is made available to the property and the time when the property connects to the utility's facilities and starts using the service.

betterment An addition or change that is made to a fixed asset and is expected to prolong that asset's life or to increase its efficiency to an extent greater than normal maintenance would accomplish. The cost of the betterment is added to the book value of the asset.

bond A written promise to pay a specified sum of money, called the face value or principal amount, at a specified date or dates in the future, called the maturity date(s), together with periodic interest at a specified rate.

bond counsel An attorney who provides expertise in securities law.

bond discount The amount of the face value of a bond in excess of the price for which the bond is acquired or sold. The price does not include accrued interest at the date of acquisition or sale.

bond indenture A document that specifies the legal obligation of the bond issuer. Typically, such an instrument contains pledges to bondholders regarding payment of principal and interest, operation, flow of funds, further debt issuance, what will constitute a default and remedy available in the event of default.

bond ordinance *See* bond indenture.

bond premium The amount by which a bond is acquired or sold in excess of its face value. The price does not include accrued interest at the date of acquisition or sale.

bond resolution *See* bond indenture.

bonded debt That portion of indebtedness represented by outstanding bonds. *See also* net bonded debt.

call (1) The process of redeeming a bond or preferred stock issue before its normal maturity. (2) An option to buy (or call) a share of stock at a specified price within a specified period.

capital budget A plan that covers all additions to and retirements from the existing utility plant.

certificate of participation A security evidencing an interest in rental payments made under a lease arrangement. In most cases the lessee (the local government) determines the lease provisions subject only to market acceptance and rating agency approval because the lessor is an entity created by the lessee in order to sell the securities.

commercial bank A financial institution that accepts deposits and offers loans. Commercial banks may offer tax-exempt loans or credit facilities to traditional municipal issuers.

competitive sale A form of bond sale in which the issuer publishes a notice of sale that specifies among other things when bids are due and how they will be evaluated.

connection charge A fee assessed by a utility to recover the cost of connecting a customer's service line to the utility's facilities.

construction work in progress (CWIP) A utility's investment in facilities that are under construction but are not yet dedicated to service. Until that property is placed in service, some regulatory agencies exclude this investment from the rate base on which utilities are permitted to earn a return.

consulting engineer A professional engineer who is engaged for a specific purpose. With respect to a debt issue, a consulting engineer generally reviews the operations and capital plans and provides a report that is included in the official statement.

contribution in aid of construction An amount of money, services, or property that is received by a water utility from any person, governmental agency, or other entity and that is provided at no cost to the utility. It represents an addition or transfer to the capital of the utility and is used to offset the acquisition, improvement, or construction costs of the utility's property, facilities, or equipment used in providing utility services to the public. It includes amounts transferred from advances for construction representing any unrefunded balances of expired refund contracts or discounts resulting from termination of refund contracts. Contributions received from governmental agencies and others for relocation of water mains or other plant facilities are also included.

credit enhancer An entity that insures bonds or provides a letter of credit to back bonds.

CWIP *See* construction work in progress.

debt service reserve fund A fund established to cover debt service for some period of time in the event of a revenue shortage, often 1-year's debt service.

deferred taxes Taxes that are charged against current operating results or reflected in the income statement and that are not currently payable to the government. Cash generated by deferring taxes is a source of funding for capital improvements.

depreciation An indication of the amount of service value not restored by current maintenance of depreciable utility plant facilities. Among the causes of this loss in value are wear and tear, decay, action of the elements, inadequacy, obsolescence, changes in demand, and requirements of public authorities.

disclosure counsel An attorney retained by the issuer to conduct due diligence, draft the official statement, and provide guidance with respect to ongoing disclosure responsibilities.

double-barrel bonds Revenue bonds backed by a specific source of revenues to which the government adds a general obligation pledge. These bonds are sometimes called combination bonds.

financial feasibility consultant A financial professional engaged to issue a report with respect to the financial feasibility of either the utility as a whole or a specific project.

floating-rate monthly demand note A note sold with no scheduled principal amortization before maturity and subject to optional redemption at par by the borrower at any monthly interest payment date.

financial printer A printer that specializes in the printing of official statements, notices of sale, and other items of a financial nature.

general obligation (GO) bonds A debt obligation issued by a government entity such as a state or municipality and backed by a pledge of the entity's full faith and credit, including its taxing authority.

GO bonds *See* general obligation bonds.

integrated resource planning (IRP) A process that maximizes available resources by considering a wide range of supply-side and demand-side resources.

investor-owned water utility A utility owned by an individual, partnership, corporation, or other qualified entity, with the equity provided by shareholders. Investor-owned utilities are generally subject to some form of regulation.

IRP *See* integrated resource planning.

issuer A borrower; in this manual, generally the water utility. This entity determines the need for bonding and selects the financing team.

issuer's counsel An attorney who represents the issuer and who reviews local legal matters for inclusion in the official statement.

lease An agreement in which the owner of property (lesssor) permits another party (lessee) to use the property in exchange for an agreed-upon payment.

letter of credit (LOC) An agreement by a bank to make a payment to the beneficiary if certain documents are presented to the bank. When the agreement obligates the bank to make principal and interest payments on bonds if the issuer is in default, a letter of credit can allow the issuer to obtain a higher credit rating.

leveraged lease A lease in which the lessor borrows the funds to acquire the property under lease and the lease payments are used to retire the debt.

line of credit An arrangement between a lender, usually a bank, and a borrower to make funds available as needed.

LOC *See* letter of credit.

long-term debt For financial statement purposes, debt having a maturity in excess of 1 year. Debt with a maturity of 2 to 10 years is often called medium-term debt.

negotiated sale A bond sale in which the issuer negotiates a price directly with the underwriter rather than selling the bonds on the market.

net bonded debt Bonded debt less any cash or other assets that are available and earmarked for that debt's retirement.

official statement A document that contains all information that a potential investor is expected to need about a security and the issuer. The preliminary official statement is issued before the sale; the final official statement is issued within 10 days after the sale and includes the prices at which the securities were offered to the public.

original-issue discount (OID) bonds Long-term bonds sold at a price below par because they have an interest rate substantially lower than prevailing market rates.

paying agent An entity, often a bank trust department, that is engaged by an issuer to disperse principal and interest to bondholders.

POS *See* preliminary official statement.

preliminary official statement (POS) A document issued before the sale that contains all information that a potential investor is expected to need about a security and the issuer, except the prices. *See* official statement.

private placement A method of sale in which the issuer sells bonds directly to a small number of sophisticated investors.

publicly owned water utility A water utility created by legislative action of a state or other government agency. A government-owned water utility may be part of a municipal government operation, a county agency, or a regional authority, or it may take such other forms as are appropriate for its service area.

put The right of an investor to require an issuer to repurchase bonds at a predetermined time and price.

put bonds *See* tender-option bonds.

rate base The value of a water utility's property as computed under any applicable laws or regulatory policies of the agency setting rates for the utility.

rate covenant A requirement stating that an issuer must pledge in a bond document to set rates sufficient to meet all operating costs and some multiple of debt service.

rating agency An independent organization that publishes, in the form of a rating, its opinion about the creditworthiness of a security.

registrar An organization, often a bank trust department, engaged by the issuer to maintain records of bond ownership.

revenue bonds A bond secured by and payable exclusively from revenues received from system operations or from the project being financed.

revenue pledge An obligation to use specified revenues for debt service.

SRF *See* State Revolving Fund.

State Revolving Fund (SRF) A program typically administered by a state department of environmental quality or health to provide loans to water utilities for capital improvements at interest rates at or below market rates.

system development charge A contribution of capital required from new utility customers or existing customers requesting enlarged or expanded services and applied toward existing or planned plant facilities necessary to meet the service needs of those customers. These charges are intended to provide funds to be used to finance all or part of capital improvements necessary to serve new customers. These charges may also be called impact fees, plant investment fees, and capital recovery charges.

tax-exempt commercial paper (TECP) A short-term promissory note with an average maturity of 30 to 45 days. Such notes, which have a maximum maturity of 270 days, are intended to be refinanced (rolled over) continuously for periods that may exceed 1 year.

TECF *See* tax-exempt commercial paper.

tender-option bonds Long-term bonds that provide the investor with the option of requiring the issuer to repurchase the bonds, generally at par, at a specified date or dates in advance of the stated maturity. Often called put bonds.

trustee An organization, generally a bank trust department, that holds funds and is responsible for protecting investors' rights.

underwriter An entity, generally an investment banking firm, that purchases bonds from an issuer and resells them to investors.

underwriter's counsel An attorney who represents the interests of the underwriter and is responsible for document preparation and review. Underwriter's counsel prepares a purchase contract between the issuer and the underwriter, as well as various underwriting documents and securities filings that are of concern only to the underwriter. It may also prepare the official statement.

user charges Fees or rates or both that are paid by customers of the system for water service.

zero-coupon bonds Long-term bonds that pay no interest before maturity. They are originally sold at a substantial discount from par.

References and Additional Sources of Information

American Telephone and Telegraph. 1977. *A Manager's Guide to Economic Decision Making*. 3rd ed. 1977. New York: McGraw-Hill.

AWWA (American Water Works Association). 1986. Water Meters—Selection, Installation, Testing, and Maintenance. Manual M6. 3rd ed. Denver, Colo.: American Water Works Association.

———. 1991. *Water Rates*. Manual M1. 4th ed. Denver, Colo.: American Water Works Association.

———. 1992. *Alternative Rates*. Manual M34. Denver, Colo.: American Water Works Association.

———. 1995. *Water Utility Accounting*. 3rd ed. Denver, Colo.: American Water Works Association.

———. 1996a. *Construction Contract Administration*. Manual M47. Denver, Colo.: American Water Works Association.

———. 1996b. *Water Rates and Related Charges*. Manual M26. 2nd ed. Denver, Colo.: American Water Works Association.

AWWARF (American Water Works Association Research Foundation). 1993. *Meeting Future Financing Needs of Water Utilities*. Denver, Colo: American Water Works Association Research Foundation and American Water Works Association.

———. 1994. *Integrated Resource Planning: A Balanced Approach to Water Resources Decision Making*. Denver, Colo.: American Water Works Association Research Foundation and American Water Works Association.

AWWA and NRRI (American Water Works Association and National Regulatory Research Institute). 1991. *Cost Allocation and Rate Design for Water Utilities*. Denver, Colo.: American Water Works Association.

Brigham, E.F. 1982. *Financial Management Theory and Practice*. 3rd ed. Chicago: Dryden Press.

Standard & Poor's Municipal Finance Criteria. New York: Standard & Poor's.

International City Manager's Association. *Evaluating Financial Condition, A Handbook for Local Government.*

Governmental Accounting Standards Board (GASB) of the Financial Accounting Foundation. 1994. *Concepts Statement No. 2 on Concepts Related to Service Efforts and Accomplishments Reporting*. Norwalk, Conn.

Kent County Water Authority Capital Improvement Program. Providence, R.I.: Camp Dresser and McKee

Lamb, R., and S.P. Rappaport. 1987. *Municipal Bond$*. 2nd ed. New York: McGraw-Hill.

Medians, Selected Indicators of Municipal Performance. 1996. New York: Moody's Investors Services, Public Finance Department. (212) 553-0470. 99 Church Street, NY 10007.

Planning for "Texas Tomorrow." 1993. Austin, Texas: Governor's Office of Budget and Planning. Unpublished.

Abbreviations

BAN	bond anticipation note
BPA	bond purchase agreement
CIP	capital improvement program
COP	certificate of participation
CP	commercial paper
CWIP	construction work in progress
GAN	grant anticipation note
GASB	Governmental Accounting Standards Board
GO bond	general obligation bond
IRP	integrated resource planning
IRS	Internal Revenue Service
LOC	letter of credit
MTN	medium-term note
NRMSIR	Nationally Recognized Municipal Securities Information Repository
OID bond	original-issue discount bond
POS	preliminary official statement
RAN	revenue anticipation note
SDC	system development charge
SEC	Securities and Exchange Commission
S&P's	Standard & Poor's Corporation
TAN	tax anticipation note
TECP	tax-exempt commercial paper

Index

NOTE: An *f.* following a page number refers to a figure. A *t.* refers to a table.

AWWA List of Manuals

M1, *Water Rates*, Fourth Edition, 1991, #30001PA

M2, *Automation and Instrumentation*, Second Edition, 1983, #30002PA

M3, *Safety Practices for Water Utilities*, Fifth Edition, 1990, #30003PA

M4, *Water Fluoridation Principles and Practices*, Fourth Edition, 1995, #30004PA

M5, *Water Utility Management Practices*, First Edition, 1980, #30005PA

M6, *Water Meters—Selection, Installation, Testing, and Maintenance*, First Edition, 1986, #30006PA

M7, *Problem Organisms in Water: Identification and Treatment*, Second Edition, 1995, #30007PA

M9, *Concrete Pressure Pipe*, Second Edition, 1995, #30009PA

M11, *Steel Pipe—A Guide for Design and Installation*, Fourth Edition, 1989, #30011PA

M12, *Simplified Procedures for Water Examination*, Second Edition, 1997, #30012PA

M14, *Recommended Practice for Backflow Prevention and Cross-Connection Control*, Second Edition, 1990, #30014PA

M17, *Installation, Field Testing, and Maintenance of Fire Hydrants*, Third Edition, 1989, #30017PA

M19, *Emergency Planning for Water Utility Management*, Third Edition, 1994, #30019PA

M20, *Water Chlorination Principles and Practices*, First Edition, 1973, #30020PA

M21, *Groundwater*, Second Edition, 1989, #30021PA

M22, *Sizing Water Service Lines and Meters*, First Edition, 1975, #30022PA

M23, *PVC Pipe—Design and Installation*, First Edition, 1980, #30023PA

M24, *Dual Water Systems*, Second Edition, 1994, #30024PA

M25, *Flexible-Membrane Covers and Linings for Potable-Water Reservoirs*, Second Edition, 1996, #30025PA

M26, *Water Rates and Related Charges*, Second Edition, 1996, #30026PA

M27, *External Corrosion—Introduction to Chemistry and Control*, First Edition, 1987, #30027PA

M28, *Cleaning and Lining Water Mains*, First Edition, 1987, #30028PA

M29, *Water Utility Capital Financing*, Second Edition, 1998, #30029PA

M30, *Precoat Filtration*, Second Edition, 1995, #30030PA

M31, *Distribution System Requirements for Fire Protection*, Second Edition, 1992, #30031PA

M32, *Distribution Network Analysis for Water Utilities*, First Edition, 1989, #30032PA

M33, *Flowmeters in Water Supply*, Second Edition, 1997, #30033PA

M34, *Alternative Rates*, First Edition, 1992, #30034PA

M35, *Revenue Requirements*, First Edition, 1990, #30035PA

M36, *Water Audits and Leak Detection*, First Edition, 1990, #30036PA

M37, *Operational Control of Coagulation and Filtration Processes*, First Edition, 1992, #30037PA

M38, *Electrodialysis and Electrodialysis Reversal*, First Edition, 1995, #30038PA

M41, *Ductile-Iron Pipe and Fittings*, First Edition, 1996, #30041PA

M44, *Valve Installation and Maintenance*, First Edition, 1996, #30044PA

M45, *Fiberglass Pipe Design*, First Edition, 1996, #30045PA

M47, *Construction Contract Administration*, First Edition, 1996, #30047PA

To order any of these manuals or other AWWA publications, call the Bookstore toll-free at 1-(800)-926-7337.